Holt Mathematics

Chapter 8 Resource Book

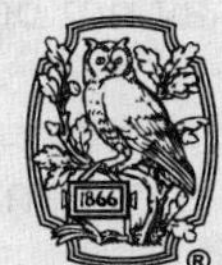

HOLT, RINEHART AND WINSTON
A Harcourt Education Company
Orlando • Austin • New York • San Diego • London

Printed in the United States of America

ISBN 0-03-078304-6

6 170 09 08

CONTENTS

Date ___________

Dear Family,

In this chapter, your child will learn how to identify, describe, and work with plane figures in geometry. The plane figures include lines and angles, as well as closed figures such as circles, polygons, triangles, and quadrilaterals. Your child will also determine whether plane figures are congruent, and will learn about symmetry. The applications in life for geometry are numerous. Your child will recognize geometry in art, sculpture, carpentry, architecture, geology, and more.

Your child will learn to use terms to describe geometric figures.

A **point** is an exact location in space.

A

point *A*
Use a capital letter to name a point.

A **line** is a set of points that extends without end in two opposite directions.

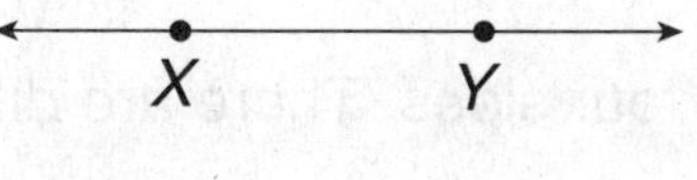

$\overleftrightarrow{XY}$
Use two points on the line to name the line.

A **ray** is part of a line.

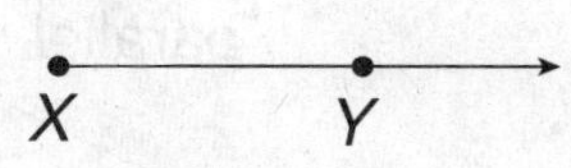

$\overrightarrow{XY}$
Name the endpoint first when naming a ray.

A **line** segment is a part of a line or a ray from one endpoint to another.

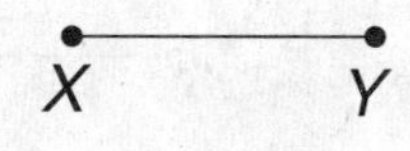

$\overline{XY}$
Use the endpoints to name a line segment.

A **plane** is a set of points that forms a perfectly flat surface that extends infinitely in all directions.

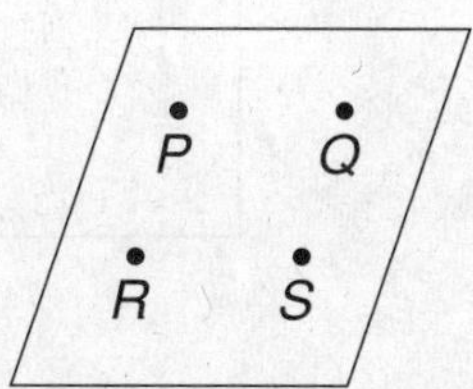

plane PQR
Use three points not on the same line to name a plane.

In learning about angles, your child will be introduced to types of angles.

A **right angle** is an angle that measures exactly 90°.

An **acute angle** is an angle that measures less than 90°.

An **obtuse angle** is an angle that measures more than 90°.

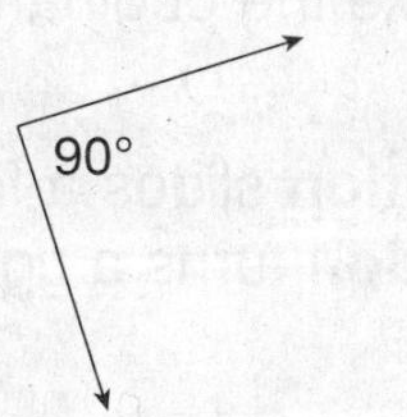

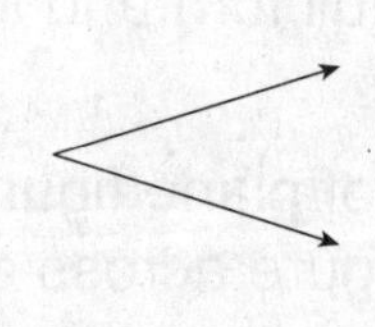

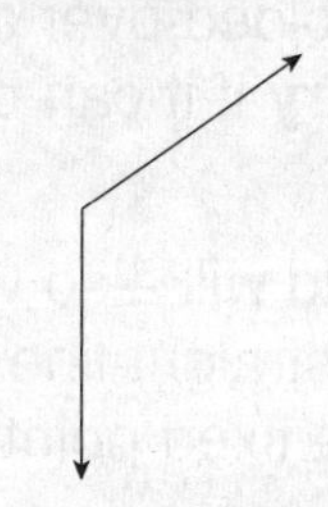

Your child will learn to identify and solve problems involving the parts of a circle.

Name all of the radii, diameters, and chords of circle *P*.

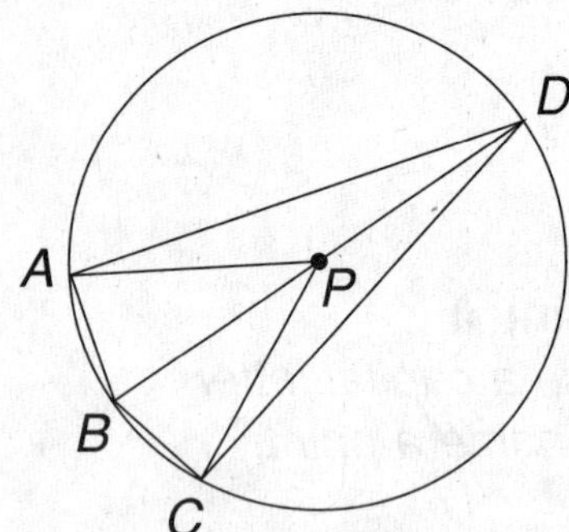

Radii: $\overline{PA}$, $\overline{PB}$, $\overline{PC}$, $\overline{PD}$

Diameter: $\overline{BD}$

Chords: $\overline{AD}$, $\overline{DC}$, $\overline{AB}$, $\overline{BC}$, $\overline{BD}$

Notice the diameter $\overline{BD}$ is also a chord and has the length of two radii.

A quadrilateral is a polygon with four sides. There are different types of **quadrilaterals.**

A **parallelogram** has two pairs of parallel sides.

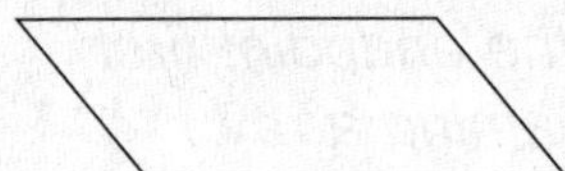

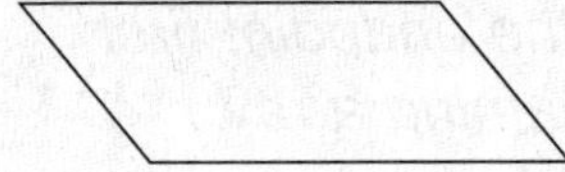

A **trapezoid** has exactly one pair of parallel sides.

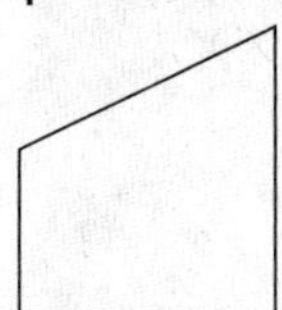

A **rhombus** has four congruent sides.

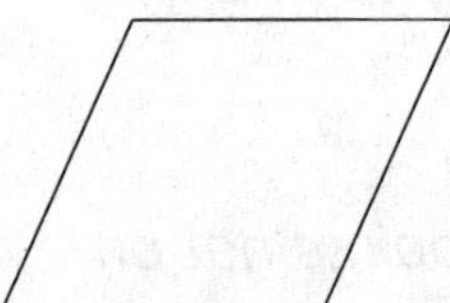

A square has four congruent sides and four right angles.

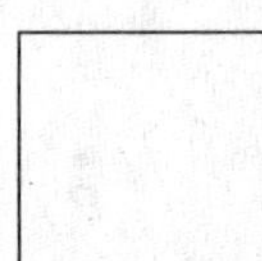

A **rectangle** has four right angles.

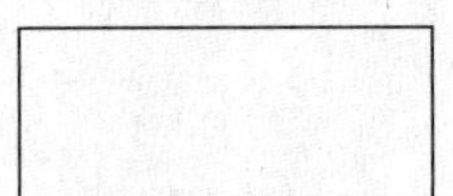

Your child will learn about congruence and symmetry. Two plane figures are **congruent** if they have the same size and the same shape. A plane figure has **line symmetry** if it can be folded over a line so that the two parts fit exactly. A plane figure has **rotational symmetry** if it can be turned less than a full turn and look exactly like the original figure.

Your child will also work with the position of plane figures. A **translation** slides a figure along a straight line. A **reflection** flips a figure across a line. A **rotation** turns a figure around a fixed point.

For additional resources, visit go.hrw.com and enter the keyword MS7 Parent.

Name ______________________________ Date ____________ Class ____________

LESSON 8-1

Practice A
Building Blocks of Geometry

Write the following in geometric notation.

1. line *EF* ______ **2.** ray *RS* ______ **3.** line segment *JK* ______

Choose the letter for the best answer.

4. Identify a line.

A $\overline{BD}$ **C** $\overleftrightarrow{CB}$

B $\overleftrightarrow{AD}$ **D** $\overrightarrow{BD}$

5. Identify a ray.

F $\overrightarrow{AC}$ **H** $\overleftrightarrow{CD}$

G $\overline{AD}$ **J** $\overrightarrow{CB}$

6. Identify a line segment.

A $\overrightarrow{DB}$ **C** $\overline{CB}$

B $\overline{DB}$ **D** $\overline{AD}$

7. Identify a plane.

F plane *AB* **H** plane *ABD*

G plane *ADE* **J** plane *BCF*

Identify the figures in the diagram.

8. two lines ______________________________

9. three rays ______________________________

10. three points ______________________________

11. three line segments ______________________________

12. a plane ______________________________

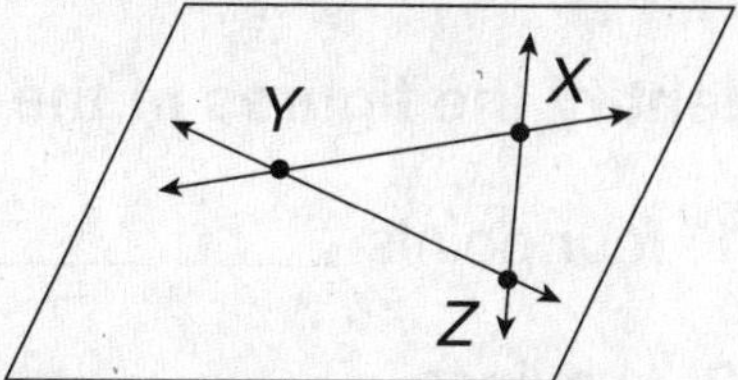

Use the figure for Exercise 13.

13. Identify which line segments are congruent.

____

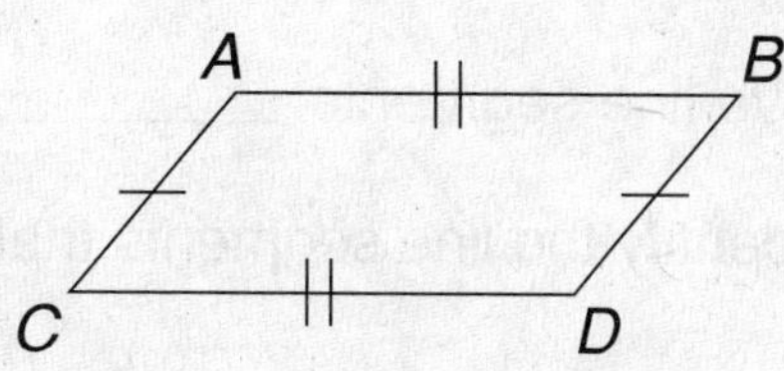

Name ______________________ Date ____________ Class ____________

LESSON 8-1

Practice B

Building Blocks of Geometry

Identify the figures in the diagram.

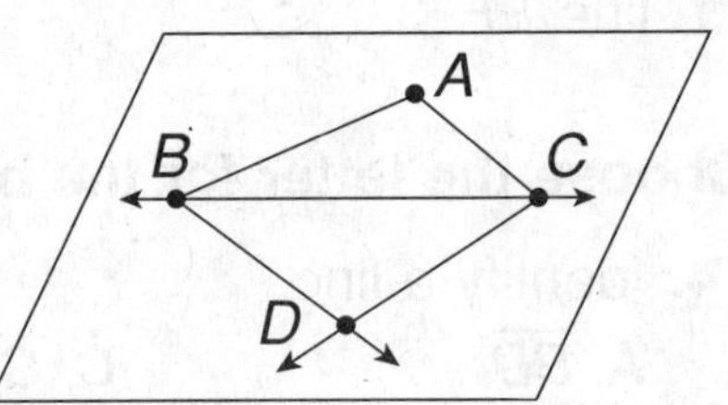

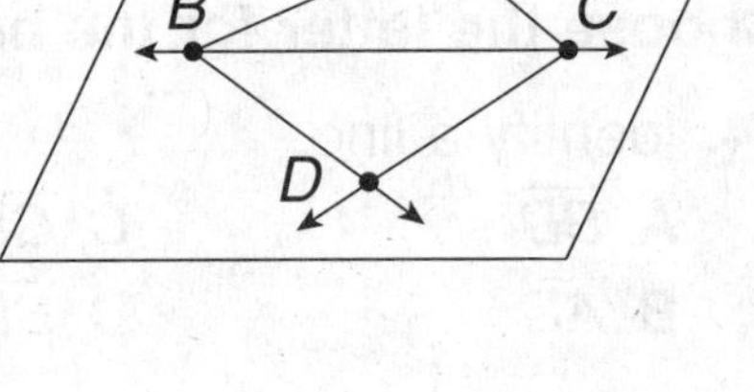

1. three points ______________________

2. one line ______________________

3. a plane ______________________

4. four rays ______________________

5. three line segments ______________________

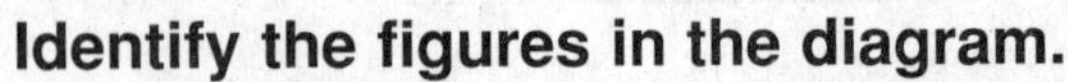

Identify the figures in the diagram.

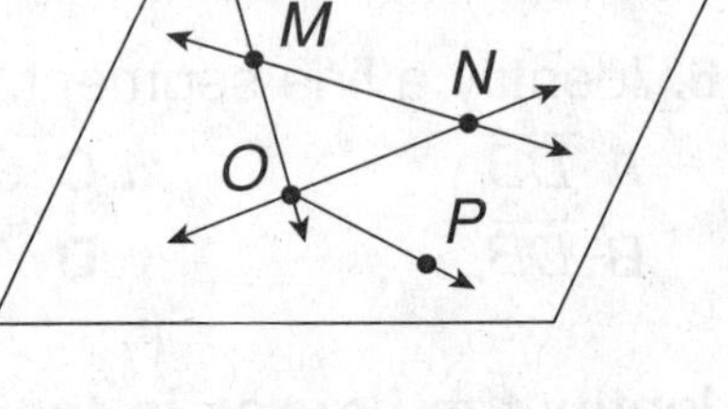

6. four points ______________________

7. three lines ______________________

8. a plane ______________________

9. three rays ______________________

10. four line segments ______________________

Identify the figures in the diagram.

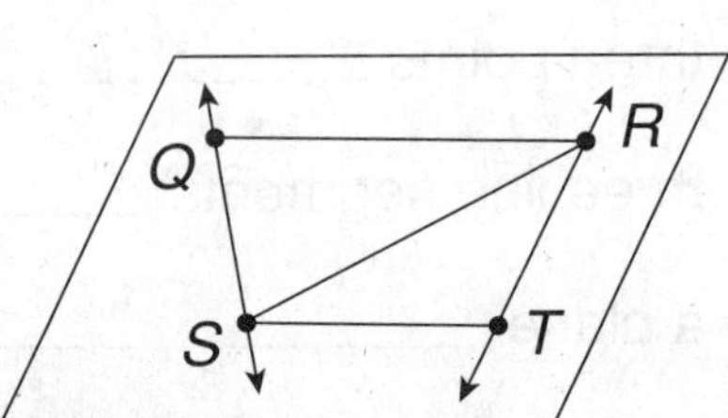

11. four points ______________________

12. two lines ______________________

13. a plane ______________________

14. four rays ______________________

15. five line segments ______________________

16. Identify the line segments that are congruent in the figure.

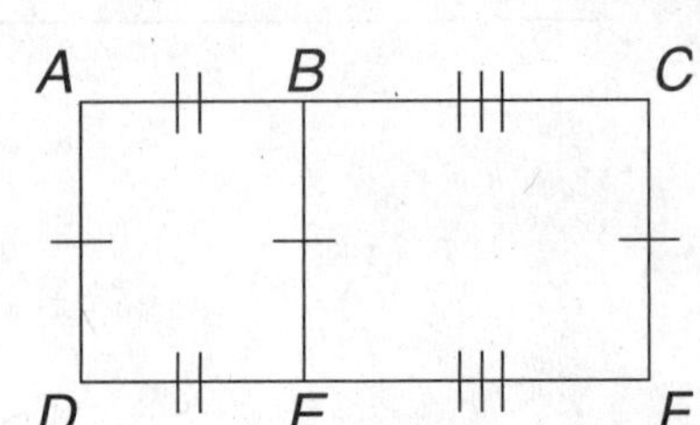

____

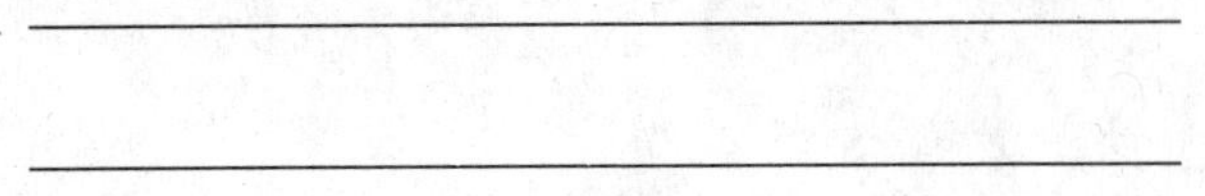

____

Name ______________________ Date ____________ Class ____________

LESSON 8-1

Practice C

Building Blocks of Geometry

Identify the figures in the diagram.

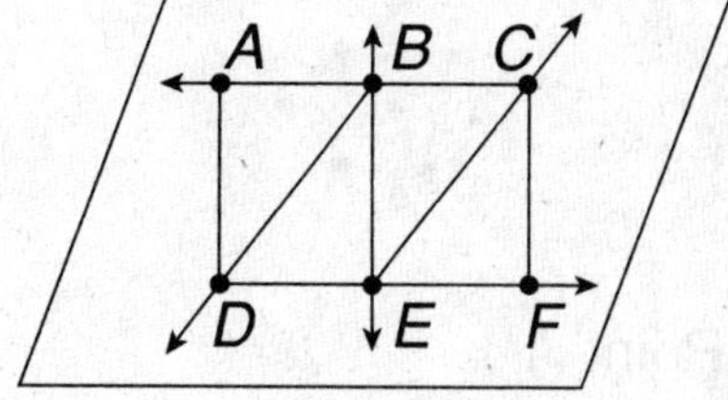

1. six points ______________________

2. a line ______________________

3. planes (list 4 possible names)

____

____

4. three rays ______________________

5. five line segments ______________________

____

Give examples of where each figure could be found in the picture.

6. points

____

7. planes

____

8. rays

____

9. line segments

____

10. congruent line segments

____

Name ____________________ Date __________ Class __________

LESSON 8-1

Reteach

Building Blocks of Geometry

You can think of real-life objects to represent geometric terms.

Point: a polka dot

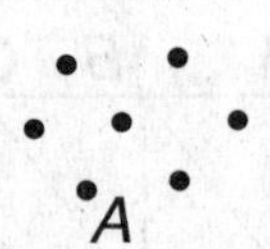

Point *A*

Line: a straight road in both directions

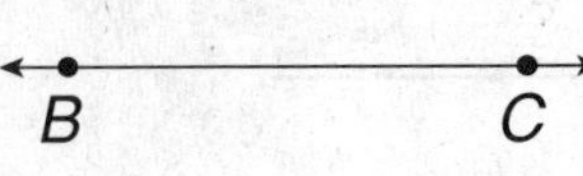

Line *BC*, or $\overleftrightarrow{BC}$

Ray: flashlight beam

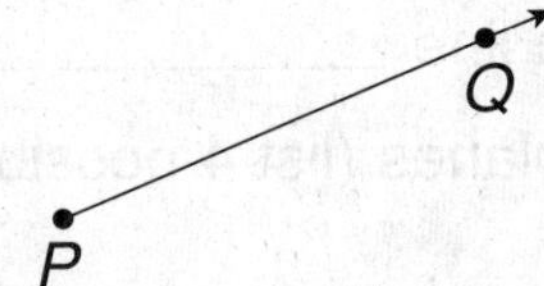

Ray *PQ*, or $\overrightarrow{PQ}$

Line segment: a ruler

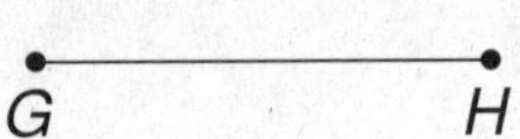

Line segment *GH*, or $\overline{GH}$

Part of a plane: table top

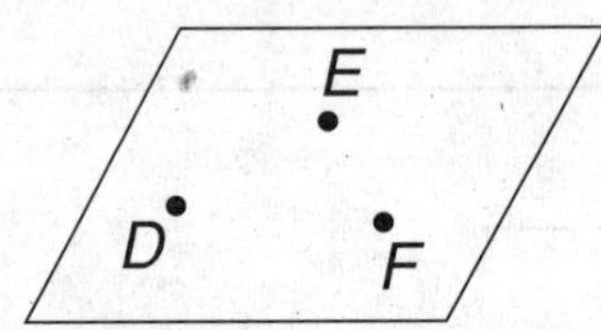

Plane *DEF*

Congruent line segments: lines on an index card

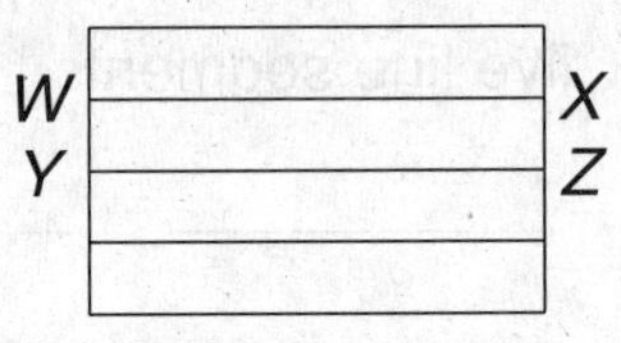

$\overline{WX} \cong \overline{YZ}$

For 1–5, use geometry notation to identify the figures.

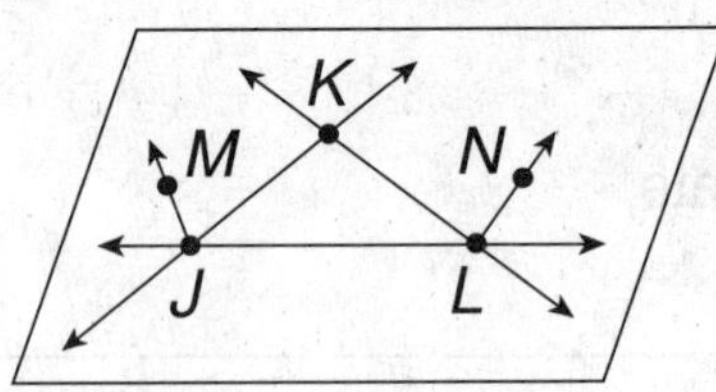

1. three points

____

2. three lines

____

3. a plane

____

4. three rays

____

5. three line segments

____

6. Identify the congruent line segments in the figure at the right.

____

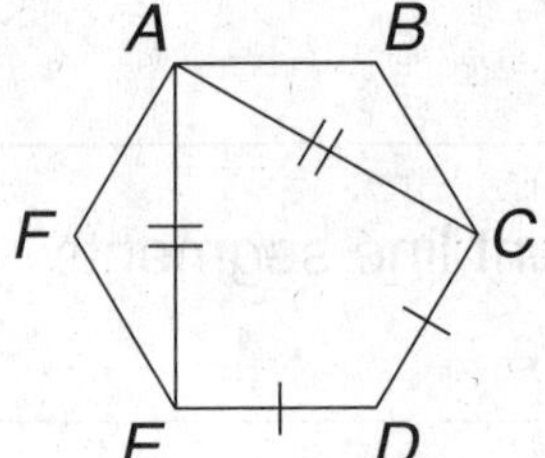

Name ______________________________ Date ____________ Class ____________

LESSON 8-1

Challenge
Plane Drawings

You can use a drawing to represent a part of a plane.

The horizontal planes below do not intersect. They are *parallel*. The dashed lines show that one plane lies behind the other plane.

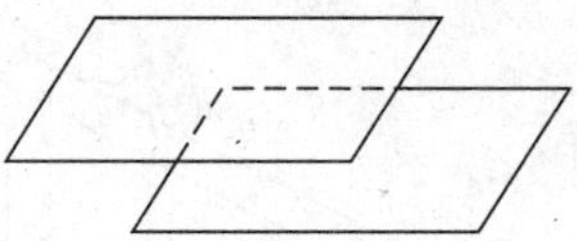

The steps below demonstrate how to sketch two intersecting planes.

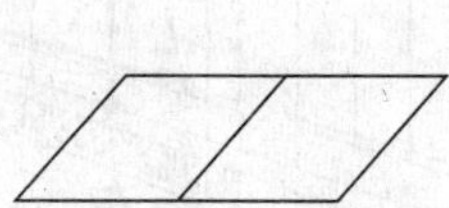

Draw a parallelogram.
Draw a line segment through its center.

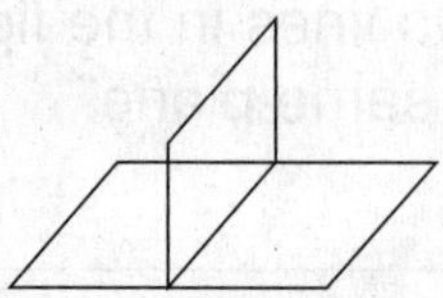

Use the segment as one side of a new parallelogram.

Draw the bottom part of a new plane.

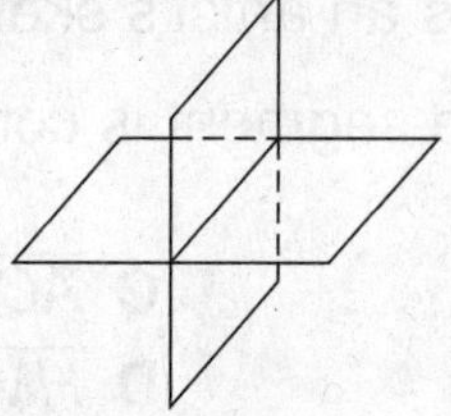

Dash the invisible edges.

Make a sketch of the planes described below.

1. Draw a pair of vertical planes that do not intersect.

2. Draw a pair of planes that intersect at an angle different from the example sketched above.

Name ______________________ Date ____________ Class ____________

LESSON **8-1**

Problem Solving

Building Blocks of Geometry

Write the correct answer.

The drawing shows a section of the Golden Gate Bridge in San Francisco.

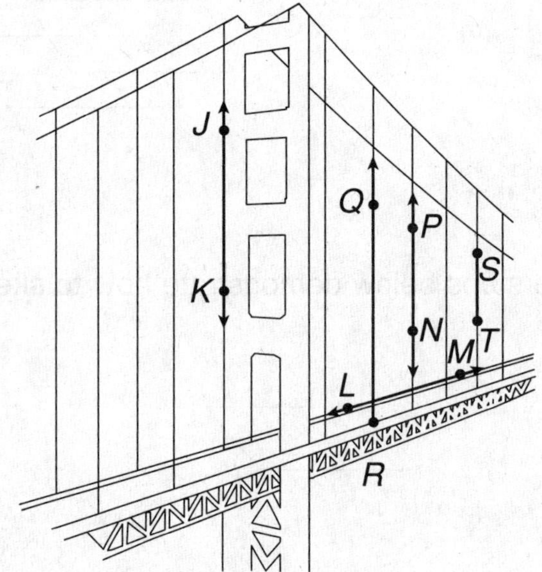

1. Identify two lines that are suggested by the bridge.

____

2. Identify a ray and a line segment that are suggested by the bridge.

____

3. Identify two lines in the figure that are in the same plane.

____

4. Identify a plane in the figure.

____

Choose the letter for the best answer.

The drawing is an artist's sketch for an abstract painting.

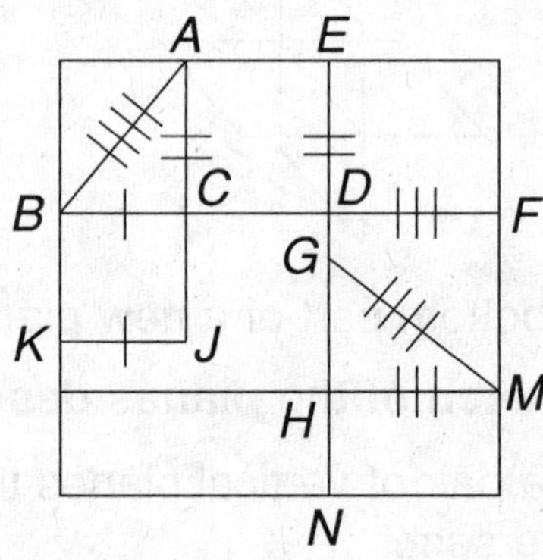

5. Which line segment is congruent to $\overline{BC}$?

A $\overline{AB}$
C $\overline{AC}$
B $\overline{KJ}$
D $\overline{HM}$

6. Which line segment is congruent to $\overline{GM}$?

F $\overline{AC}$
H $\overline{DE}$
G $\overline{DF}$
J $\overline{AB}$

7. Which line segment is congruent to $\overline{ED}$?

A $\overline{AC}$
C $\overline{DF}$
B $\overline{CD}$
D $\overline{GH}$

8. Which line segment is congruent to $\overline{DF}$?

F $\overline{ED}$
H $\overline{CJ}$
G $\overline{HM}$
J $\overline{FM}$

Name ______________________ Date __________ Class __________

LESSON 8-1

Reading Strategies
Vocabulary Development

A **point** marks an exact location in space. It is usually shown as a dot.

J • Read: point J

A **line** is a straight path that extends forever in both directions.

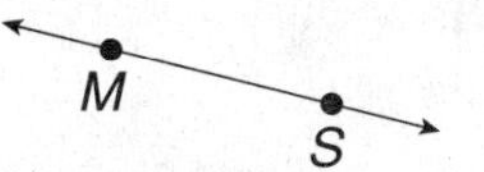

Read: line MS or $\overleftrightarrow{MS}$ or line SM or $\overleftrightarrow{SM}$

A **ray** is part of a line that has one endpoint and goes on forever in one direction.

Read: ray TD or $\overrightarrow{TD}$

A **line segment** is part of a line between two endpoints. The two endpoints are used to name the line segment.

Q, Z

Read: segment QZ or $\overline{QZ}$ or segment ZQ or $\overline{ZQ}$

Congruent line segments have the same length. Tick marks are used to show congruent segments.

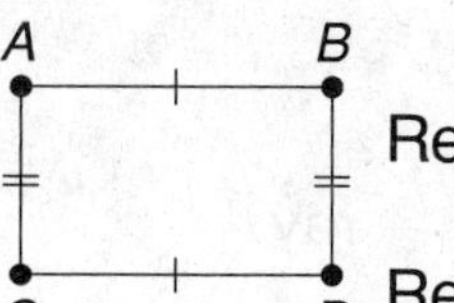

$\overline{AB} \cong \overline{CD}$

Read: $\overline{AB}$ is congruent to $\overline{CD}$

$\overline{AC} \cong \overline{BD}$

Read: $\overline{AC}$ is congruent to $\overline{BD}$

A flat surface that extends without end in all directions is called a **plane.** A plane is named by three points on the plane that are not on a line.

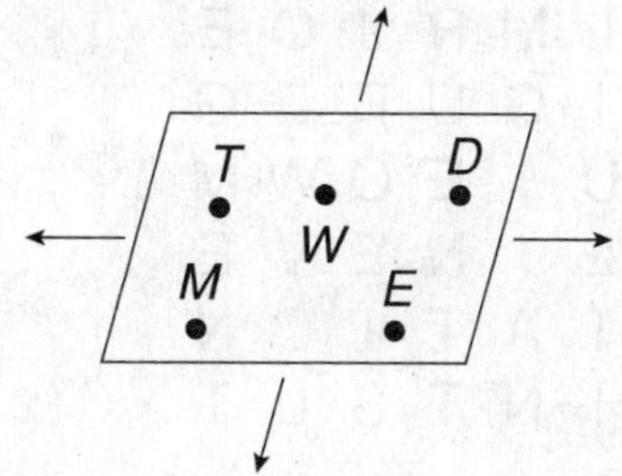

This plane could be named plane *MTW*.

Use the information above to answer each question.

1. What name is given to a path that goes on in both directions?

____

2. What name is given to part of a line with two endpoints?

____

3. What name is given to part of a line with only one endpoint?

____

4. What name is given to line segments with the same length?

____

5. How would you describe a plane?

____

Name ________________________ Date ____________ Class ____________

LESSON 8-1

Puzzles, Twisters & Teasers

Are You Barking Up the Wrong Tree?

Find and circle words from the list in the word search (horizontally, vertically or diagonally). Then find a word in the word search that solves the riddle. Circle it and write it on the line.

points	notation	ray	line	segment
plane	congruent	figure	geometric	endpoint

```
G E O M E T R I C C E R
Z N M H T G A Y O V G Y
C D E T U I Y P N B H U
V P L A N E Z A G N I S
B O I B I P L M R J O E
N I V A K F I G U R E G
M N O R B H U I E Q W M
L T M K W E L I N E V E
K I J G N O T A T I O N
L D V G P O I N T S L T
```

How can you identify a dogwood tree?

By its ___ ___ ___ ___

Name ______________________ Date __________ Class __________

LESSON 8-2

Practice A

Classifying Angles

Tell whether each angle is acute, right, obtuse, or straight.

1.

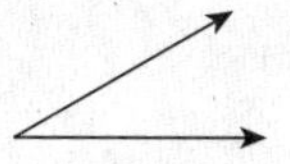

____

2.

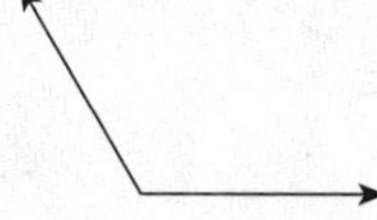

____

3.

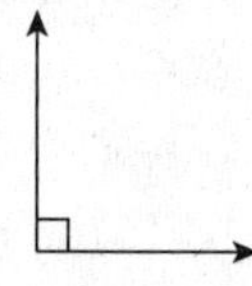

____

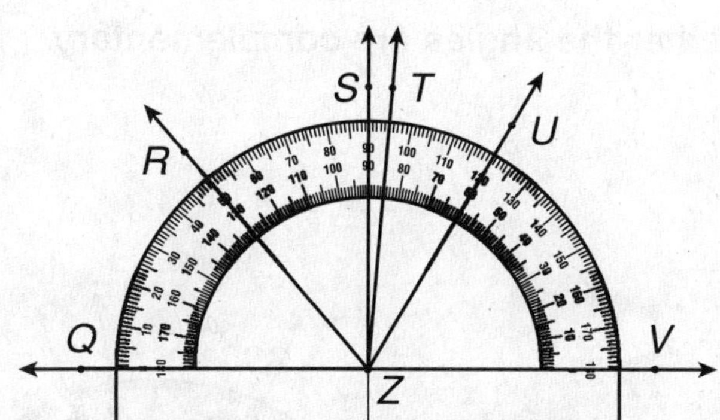

Use the diagram to complete Exercises 4-8.

4. Add the measures of $\angle UZV$ and $\angle TZU$. ______________

5. Are $\angle UZV$ and $\angle TZU$ complementary, supplementary, or neither? ______________

6. Add the measures of $\angle QZR$ and $\angle RZS$. ______________

7. Are $\angle QZR$ and $\angle RZS$ complementary, supplementary, or neither? ______________

8. Are $\angle QZS$ and $\angle VZS$ complementary, supplementary, or neither? ______________

9. Angles G and H are supplementary. If $m\angle G$ is 80°, what is $m\angle H$? ______________

10. Angles M and N are complementary. If $m\angle M$ is 20°, what is $m\angle N$? ______________

Name ______________________ Date __________ Class __________

LESSON 8-2

Practice B

Classifying Angles

Tell whether each angle is acute, right, obtuse, or straight.

1.

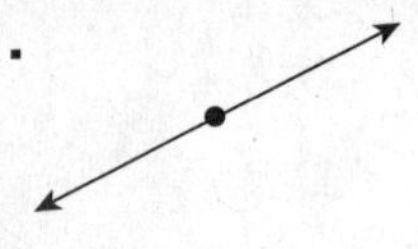

____

2.

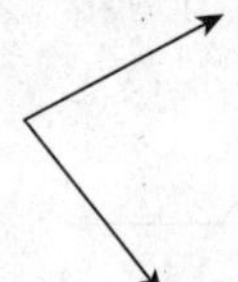

____

3.

____

Use the diagram to tell whether the angles are complementary, supplementary or neither.

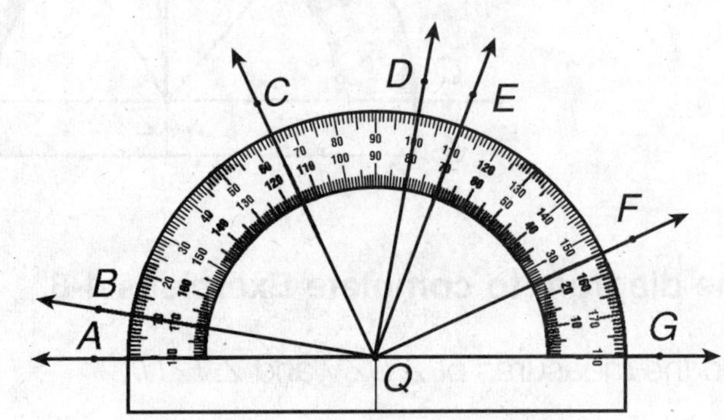

4. ∠*AQC* and ∠*GQC*

____

5. ∠*BQD* and ∠*DQE*

____

6. ∠*CQE* and ∠*EQF*

____

7. ∠*GQF* and ∠*FQE*

____

8. ∠*BQC* and ∠*DQC*

____

9. Angles *W* and *X* are supplementary. If m∠*W* is 37°, what is m∠*X*? _______________

10. Angles *S* and *T* are complementary. If m∠*S* is 64°, what is m∠*T*? _______________

11. Angles *C* and *D* are supplementary. If m∠*C* is 83°, what is m∠*D*? _______________

12. Angles *U* and *V* are complementary. If m∠*U* is 41°, what is m∠*V*? _______________

Name ______________________ Date __________ Class __________

LESSON 8-2

Practice C
Classifying Angles

Tell whether each angle is acute, right, obtuse, or straight.

1.

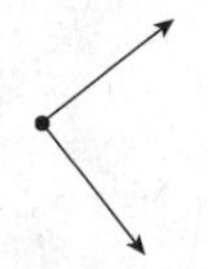

____

2.

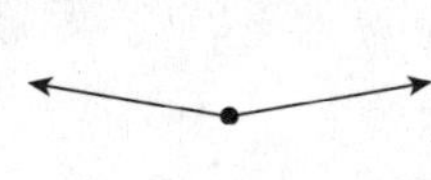

____

3.

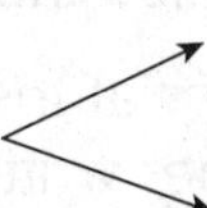

____

Use the diagram to tell whether the angles are complementary, supplementary or neither.

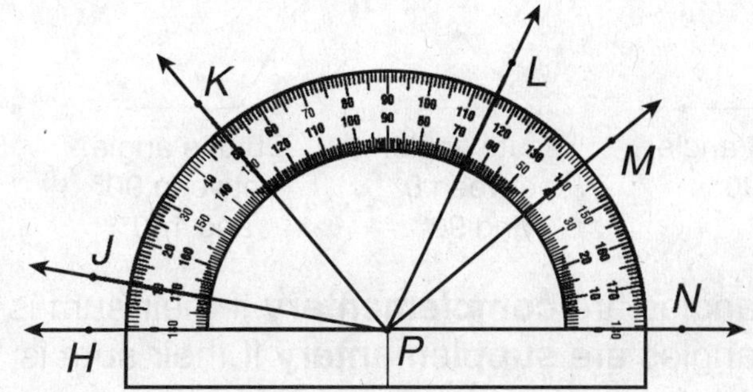

4. ∠*JPK* and ∠*KPL*

____

5. ∠*LPK* and ∠*MPL*

____

6. ∠*HPM* and ∠*MPN*

____

7. ∠*LPM* and ∠*KPJ*

____

8. Angles *L* and *M* are complementary. If m∠*L* is 16°, what is m∠*M*?

____

9. Angles *R* and *S* are supplementary. If m∠*R* is 78°, what is m∠*S*?

____

Classify each pair of angles as complementary or supplementary. Then find the missing angle measure.

10.

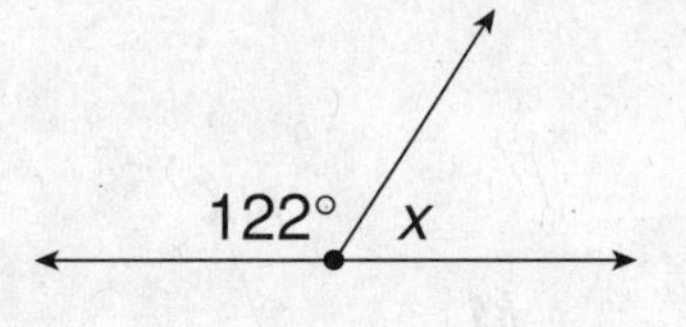

____

11.

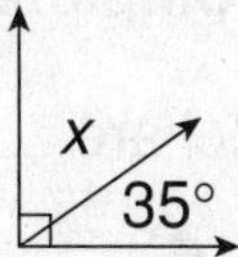

____

Name ______________________ Date __________ Class __________

LESSON 8-2

Reteach

Classifying Angles

Two rays with a common endpoint form an **angle.** The common endpoint is called the **vertex.** You can name an angle three ways:

- with the letter at the vertex: ∠*Y*.
- with a number written inside the angle: ∠1.
- with three letters: ∠*XYZ* or ∠*ZYX*. The letter of the vertex must be in the middle.

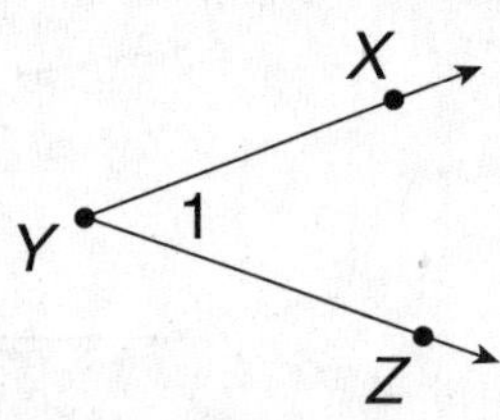

Angles are measured in degrees and classified by their measures.

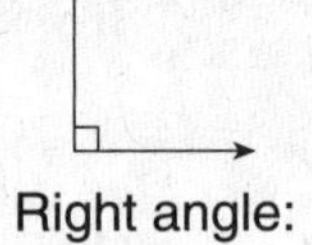

Right angle: 90°	Acute angle: between 0° and 90°	Obtuse angle: between 90° and 180°	Straight angle: 180°

Two angles are **complementary** if their sum is 90°.
Two angles are **supplementary** if their sum is 180°.

Use the figure to complete the statements.

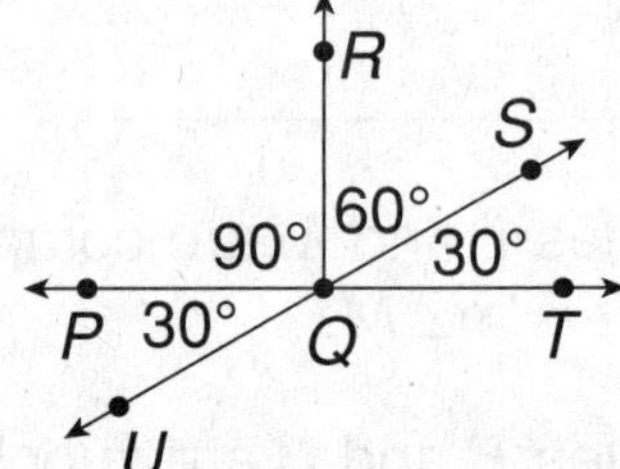

1. ∠*PQR* measures ______.
   It is a ________ angle.
2. ∠*RQS* measures ______.
   It is an ________ angle.
3. ∠*UQR* measures 30° + 90°, or ______.
   It is an ________ angle.
4. ∠*PQT* measures 90° + 60° + 30°, or ______.
   It is a ________ angle.

Use the figure to complete the statements.

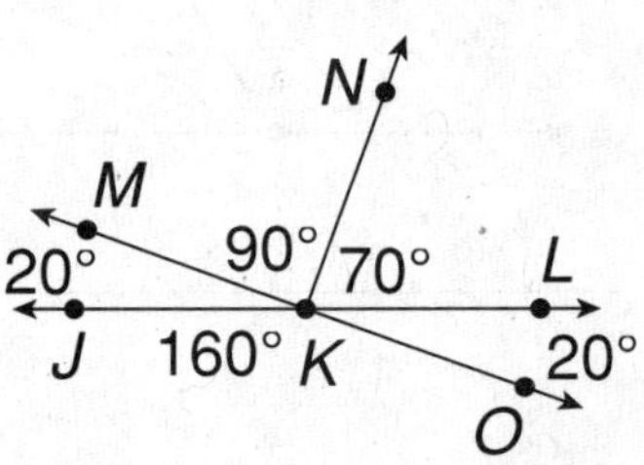

5. ∠*NKL* and ∠*LKO* are ________________ angles since 70° + 20° = ______.
6. ∠*JKO* and ∠*OKL* are ________________ angles since 160° + 20° = ______.

Name ______________________________ Date __________ Class __________

LESSON 8-2

Challenge
Angling Clocks

The hands of a clock form an angle. At 9 o'clock, the hour hand points to 9, and the minute hand points to 12. The angle formed measures 90°.

But how do you measure the angle when both hands are between numbers?

- There are 5 minutes between two consecutive numbers on a clock.
- Each time the minute hand moves 12 minutes, or $\frac{1}{5}$ the distance around the clock, the hour hand moves $\frac{1}{5}$ the distance between two consecutive numbers.

What is the angle between the hands at 7:24?

When the minute hand moves 24 minutes, the hour hand moves 2 sections. So, there are $(7 \cdot 5 + 2) - 24$, or 13, sections between the hands. Each section is $\frac{1}{60}$ of an hour, and there are 360° on a clock face.

$13 \cdot \frac{1}{60} \cdot 360 = 13 \cdot 6 = 78$

The angle between the hands measures 78°.

1. Draw the clock hands and find the angle between them at 5 o'clock.

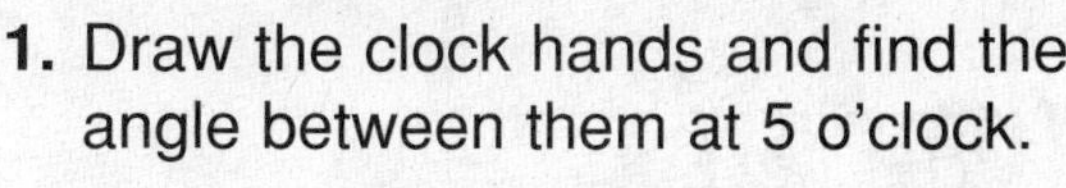

2. Draw the clock hands and find the angle between them at 8 o'clock.

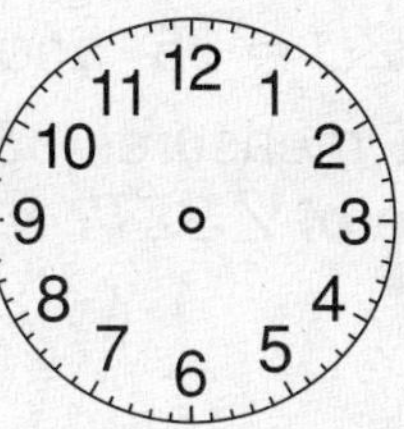

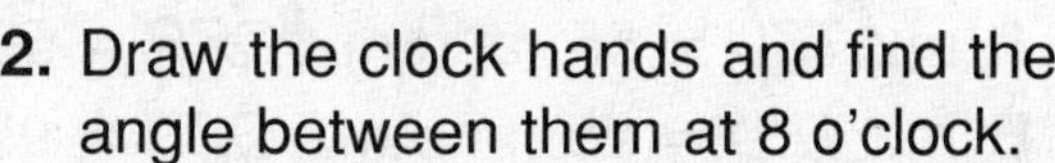

3. Find the angle between the hands at 1 o'clock. ______

4. Find the angle between the hands at 2 o'clock. ______

5. Find the angle between the hands at 12:24. ______

6. Find the angle between the hands at 3:48. ______

7. Find the angle between the hands at 3:12. ______

8. Find the angle between the hands at 1:36. ______

Name ________________________ Date ____________ Class ____________

LESSON 8-2

Problem Solving

Classifying Angles

Write the correct answer.

The drawing shows a scene on a calendar.

1. ∠1 and ∠2 are complementary angles. If ∠1 measures 35°, what is the measure of ∠2?

____

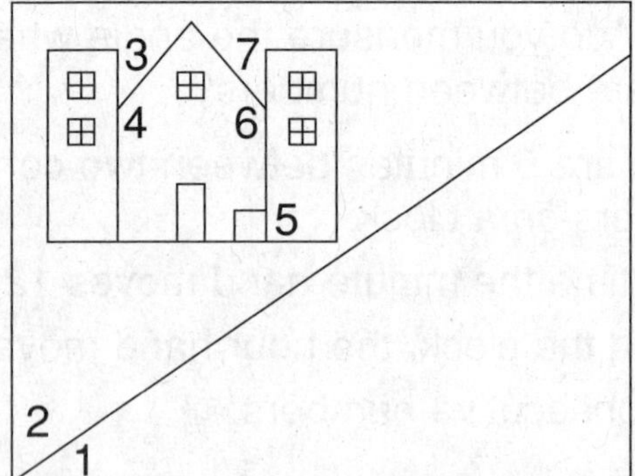

2. ∠3 and ∠4 are supplementary angles. If ∠3 measures 50°, what is the measure of ∠4?

____

3. Which angle is an obtuse angle: ∠6 or ∠7?

____

4. Which angle labeled on the drawing is a right angle?

____

Choose the letter for the correct answer.

Use the diagram to complete Exercises 5 and 6.

5. Which of the following could be the measures of ∠*TZU* and ∠*QZR*?

A m∠*TZU* = 55° and m∠*QZR* = 55°
B m∠*TZU* = 25° and m∠*QZR* = 95°
C m∠*TZU* = 80° and m ∠*QZR* = 100°
D m∠*TZU* = 35° and m∠*QZR* = 80°

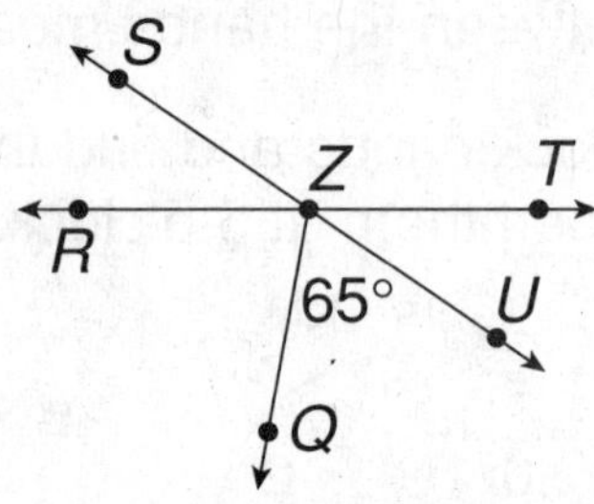

6. If ∠*RZS* measures 35°, what is the measure of ∠*SZT*?

F 155°
G 145°
H 55°
J 45°

7. ∠*A* and ∠*B* are complementary angles. The measure of ∠*B* is 4 times the measure of ∠*A*. What are the measures of the angles?

A m∠*A* = 16° and m∠*B* = 64°
B m∠*A* = 18° and m∠*B* = 72°
C m∠*A* = 36° and m ∠*B* = 144°
D m∠*A* = 45° and m∠*B* = 135°

8. The hands of a clock form an acute angle at 1:00. What type of angle do they form at 4:00?

F acute
G right
H obtuse
J straight

Name ______________________ Date ____________ Class ____________

LESSON **8-2**

Reading Strategies

Use Manipulatives

You use a ruler to measure the length of a line. You use a **protractor** to measure the degrees of an angle.

Using a protractor, follow these steps to measure an angle.

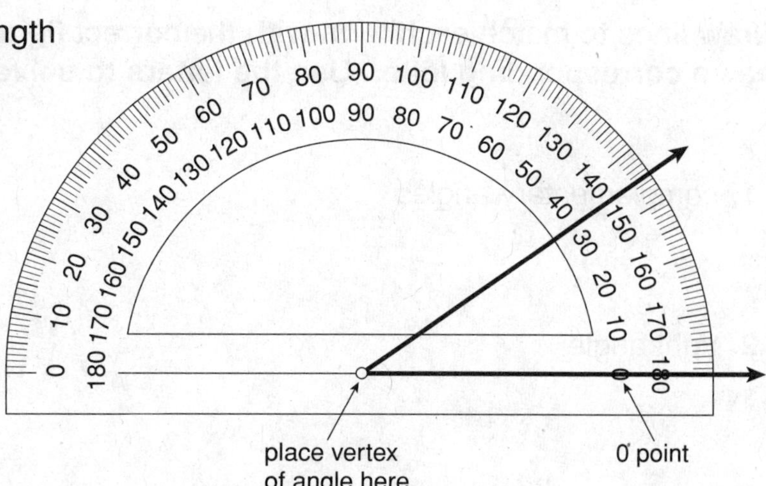

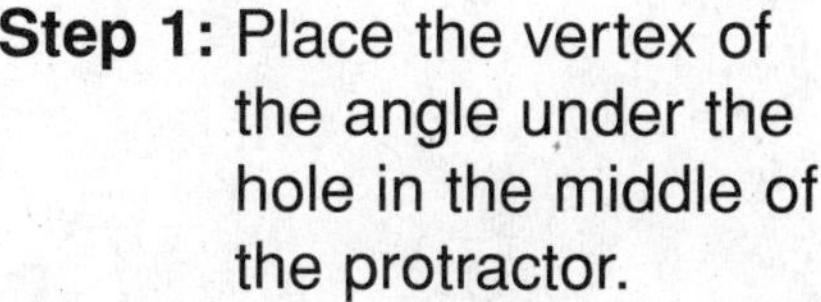

Step 1: Place the vertex of the angle under the hole in the middle of the protractor.

Step 2: Line up the 0 line on the protractor with one of the rays. This ray is lined up with the 0 line on the right side of the protractor, so you will use the inside numbers to find the measure of the angle.

Step 3: Follow the inside numbers from 0 until you reach the other ray. Notice that the other ray is on 35°. The measure of the angle is 35°.

Use the information above to answer each question.

1. What tool is used to measure an angle? ____________

2. Where is the vertex of an angle placed on your protractor to measure the angle?

____

3. What part of an angle lines up with the 0 line on the protractor?

____

Use a protractor to measure these angles.

4.

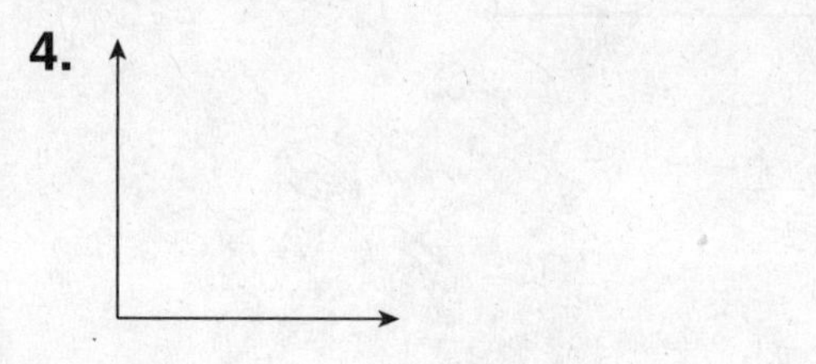

____

5.

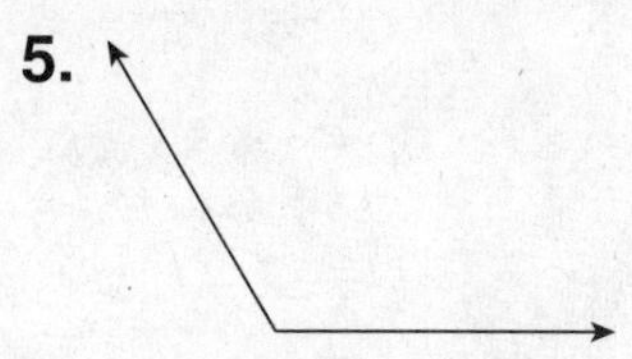

____

Name ______________________ Date ____________ Class ____________

LESSON 8-2

Puzzles, Twisters & Teasers

Off the Hook!

Draw lines to match each term with the correct figure. Each figure has a corresponding letter. Use the letters to solve the riddle.

1. complementary angles
2. right angle
3. ray
4. supplementary angles
5. vertex
6. obtuse angle
7. acute angle
8. straight angle

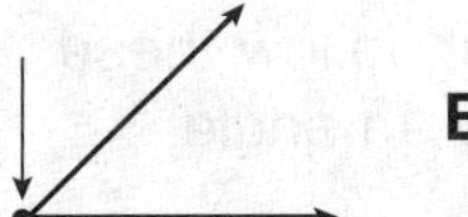
E

O

S

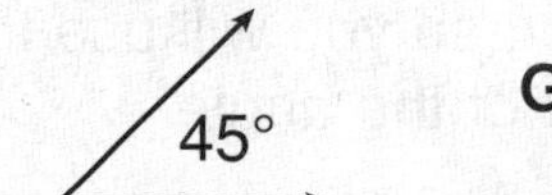

G

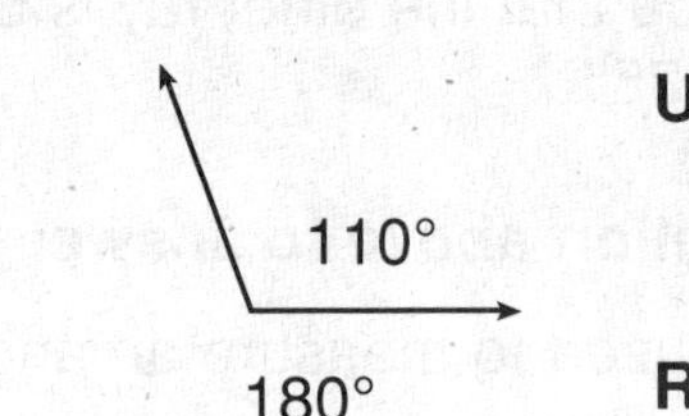

U

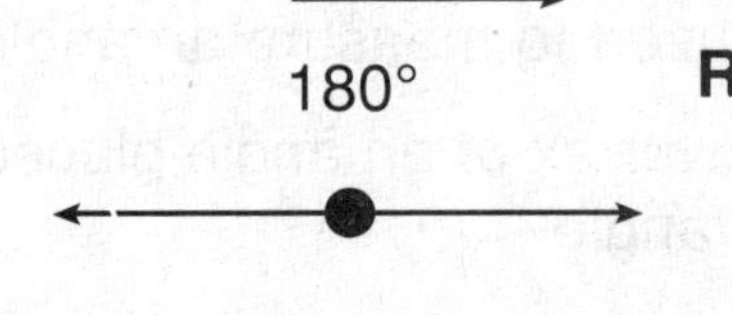

R

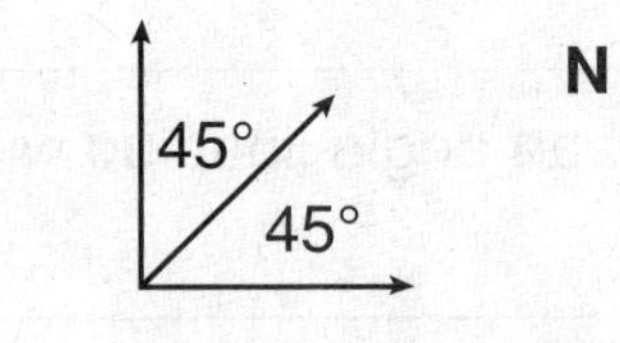

N

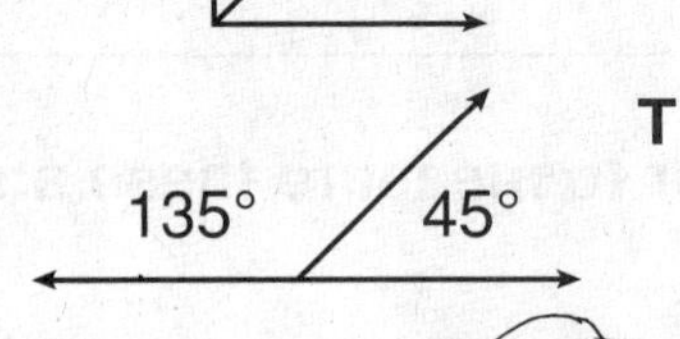

T

Who performs operations in the fish hospital?

The ____ ____ ____ ____ ____ ____ ____ ____

Name ______________________ Date ____________ Class ____________

LESSON 8-3

Practice A

Angle Relationships

1. Draw a pair of parallel lines.

2. Draw a pair of perpendicular lines.

Tell whether each statement is true or false.

3. $\overleftrightarrow{AC}$ and $\overleftrightarrow{AB}$ appear parallel. ____________

4. $\overleftrightarrow{AC}$ and $\overleftrightarrow{EF}$ appear parallel. ____________

5. $\overleftrightarrow{AB}$ and $\overleftrightarrow{CD}$ appear perpendicular. ____________

6. $\overleftrightarrow{CF}$ and $\overleftrightarrow{FG}$ appear perpendicular. ____________

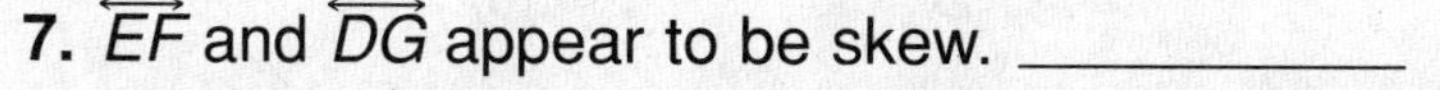

7. $\overleftrightarrow{EF}$ and $\overleftrightarrow{DG}$ appear to be skew. ____________

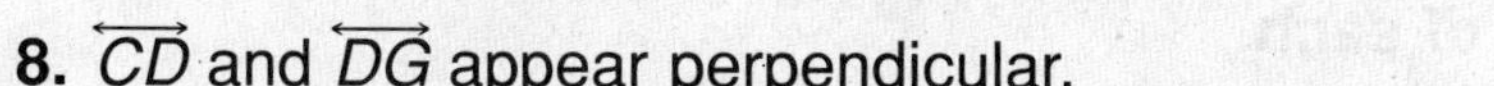

8. $\overleftrightarrow{CD}$ and $\overleftrightarrow{DG}$ appear perpendicular. ____________

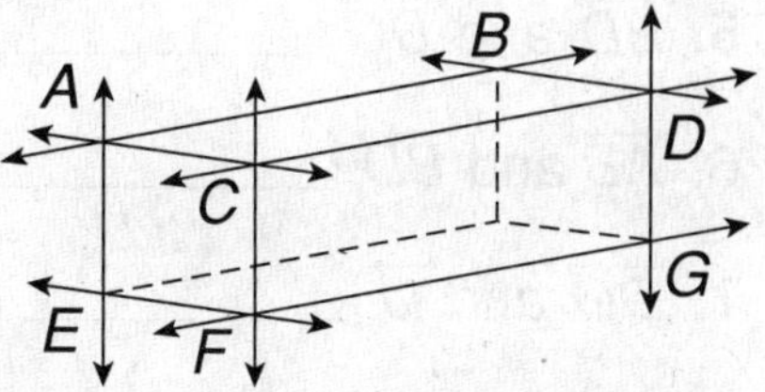

Line *x* is parallel to line *y*. Find the measure of each angle.

9. ∠1 and ∠6

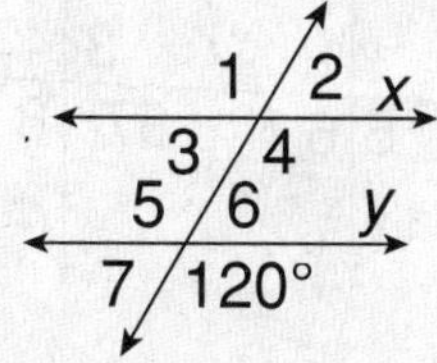

____

10. ∠3 and ∠7

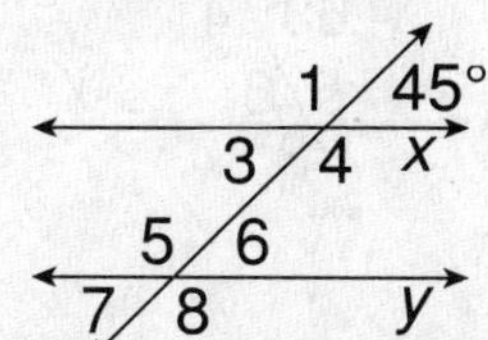

____

11. ∠1 and ∠4

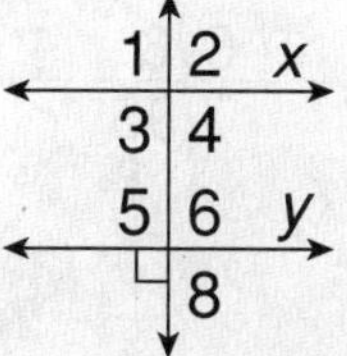

____

12. ∠6 and ∠4

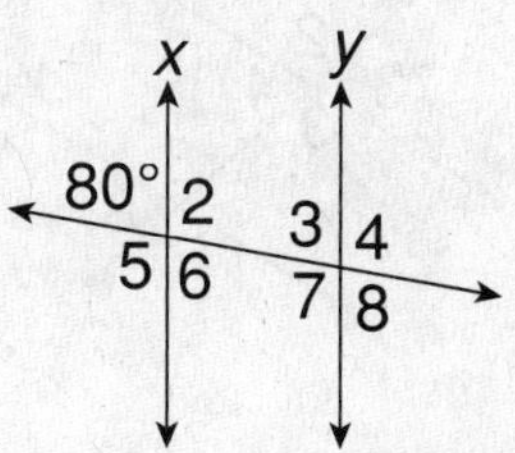

____

13. ∠6 and ∠7

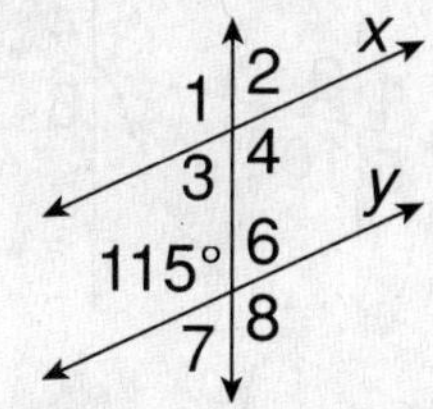

____

14. ∠1 and ∠3

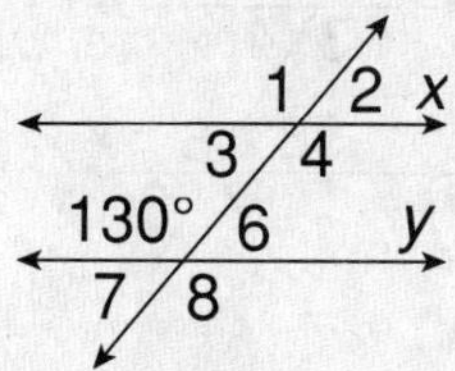

____

Name ______________________ Date ____________ Class ____________

LESSON **8-3**

Practice B

Angle Relationships

Tell whether the lines appear parallel, perpendicular, or skew.

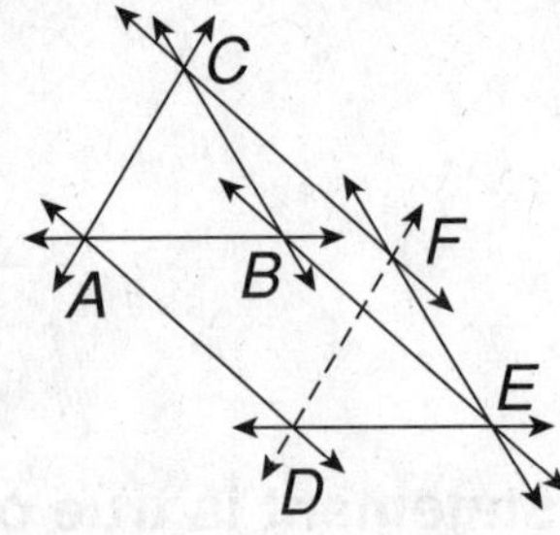

1. $\overleftrightarrow{AB}$ and $\overleftrightarrow{DE}$ ____________________

2. $\overleftrightarrow{EF}$ and $\overleftrightarrow{CF}$ ____________________

3. $\overleftrightarrow{AB}$ and $\overleftrightarrow{AD}$ ____________________

4. $\overleftrightarrow{BC}$ and $\overleftrightarrow{DE}$ ____________________

Tell whether the lines appear parallel, perpendicular, or skew.

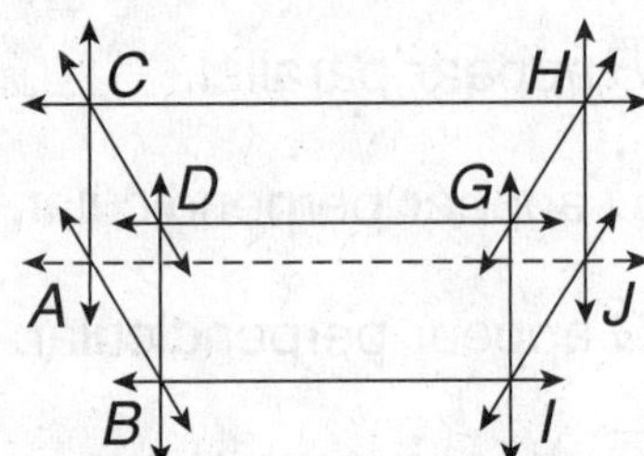

5. $\overleftrightarrow{BD}$ and $\overleftrightarrow{DG}$ ____________________

6. $\overleftrightarrow{AB}$ and $\overleftrightarrow{BD}$ ____________________

7. $\overleftrightarrow{DG}$ and $\overleftrightarrow{IJ}$ ____________________

8. $\overleftrightarrow{AB}$ and $\overleftrightarrow{CD}$ ____________________

Line $x \parallel$ line y. Find the measure of each.

9. $\angle 1$ and $\angle 6$

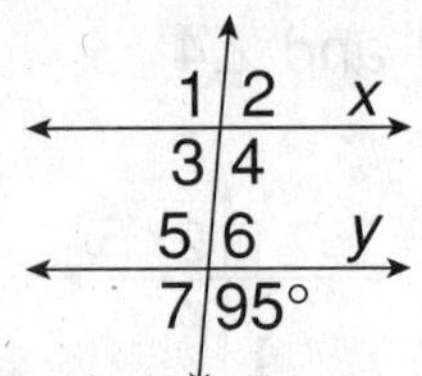

____

10. $\angle 4$ and $\angle 8$

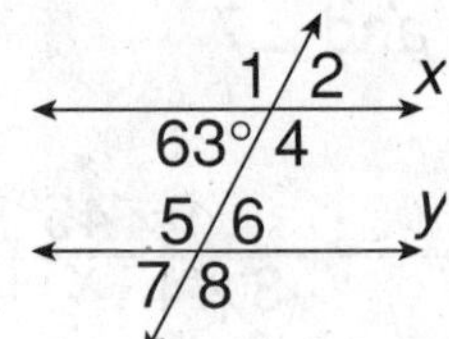

____

11. $\angle 4$ and $\angle 6$

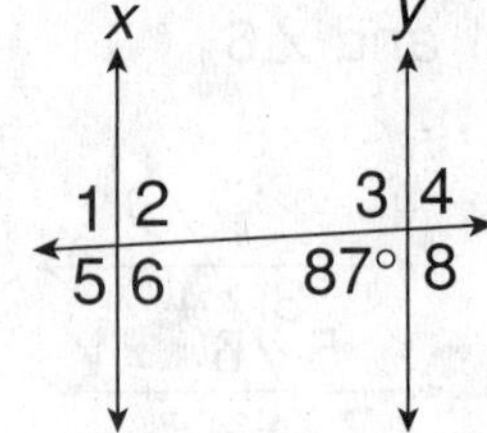

____

12. $\angle 2$ and $\angle 4$

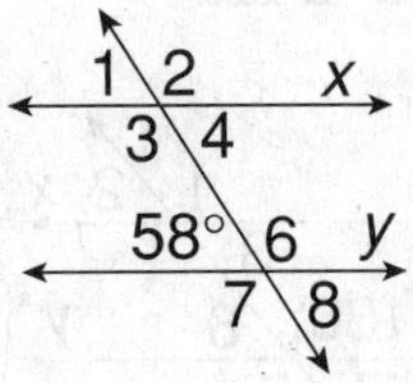

1 2 x
3 4
58° 6 y
7 8

____

13. $\angle 5$ and $\angle 7$

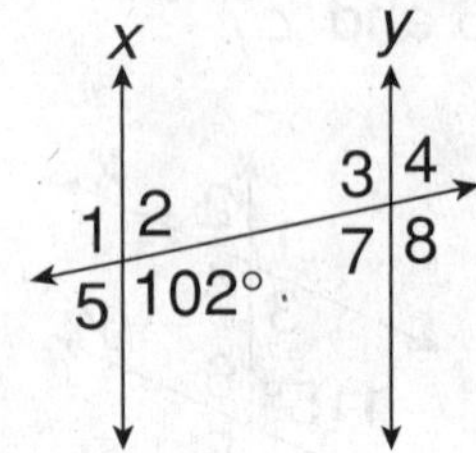

____

14. $\angle 7$ and $\angle 8$

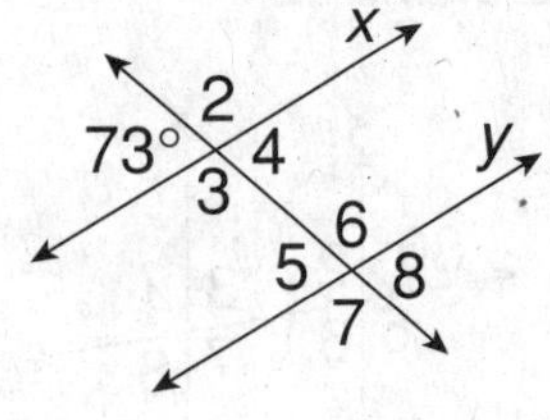

____

Name ____________________ Date __________ Class __________

LESSON **8-3**

Practice C
Angle Relationships

Tell whether the lines appear parallel, perpendicular or skew.

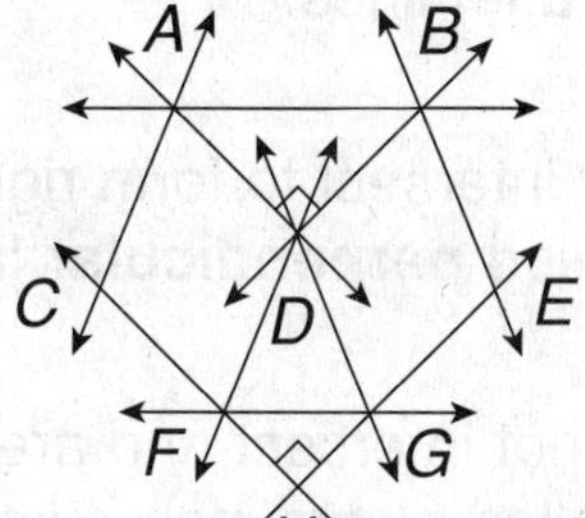

1. $\overleftrightarrow{DF}$ and $\overleftrightarrow{AC}$ ____________

2. $\overleftrightarrow{AD}$ and $\overleftrightarrow{BD}$ ____________

3. $\overleftrightarrow{AD}$ and $\overleftrightarrow{GE}$ ____________

4. $\overleftrightarrow{DG}$ and $\overleftrightarrow{BE}$ ____________

5. $\overleftrightarrow{AB}$ and $\overleftrightarrow{GH}$ ____________

6. $\overleftrightarrow{FH}$ and $\overleftrightarrow{GH}$ ____________

Assume that lines that look parallel are parallel. Find the measure of each angle.

7. ∠4 and ∠6

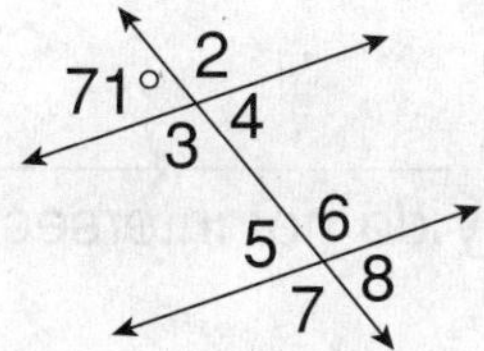

____

8. ∠1 and ∠7

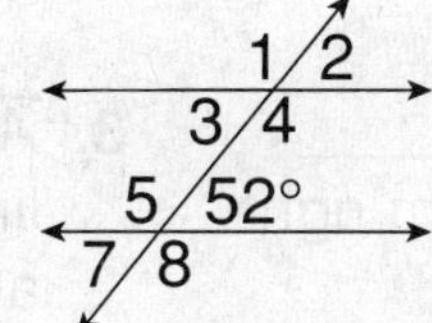

____

9. ∠3 and ∠8

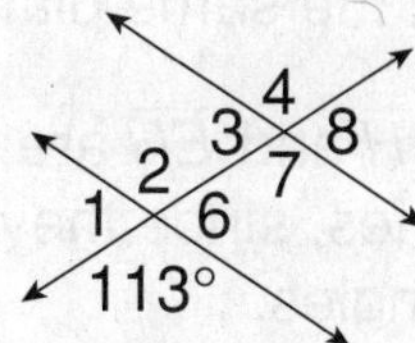

____

10. ∠2, ∠5, and ∠12

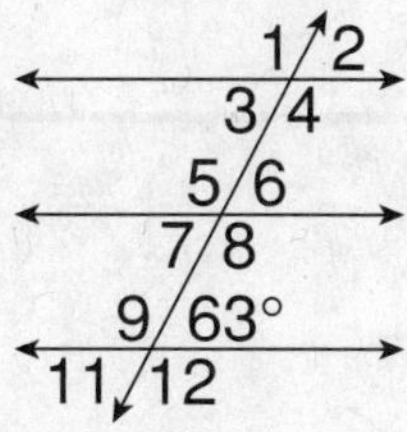

____

11. ∠1, ∠8, and ∠10

1 2
3 4
85° 6
7 8
9 10
11 12

____

12. ∠3, ∠6, and ∠11

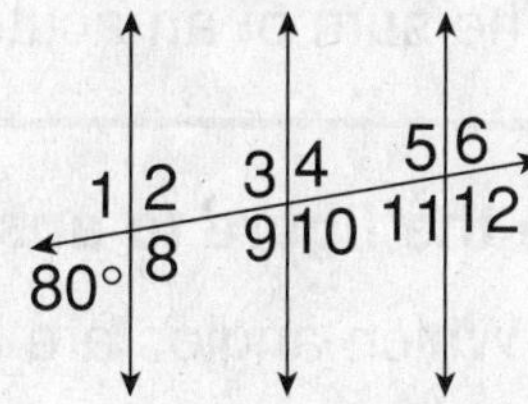

____

13. A pair of supplementary angles is congruent. What is the measure of each angle?

____

Name ______________________ Date ____________ Class ____________

LESSON 8-3

Reteach
Angle Relationships

Lines in the same plane that never meet are called **parallel** lines.

Two lines that intersect to form right angles are called **perpendicular** lines.

Lines that do not intersect and are not parallel are called **skew** lines. Skew lines are in different planes.

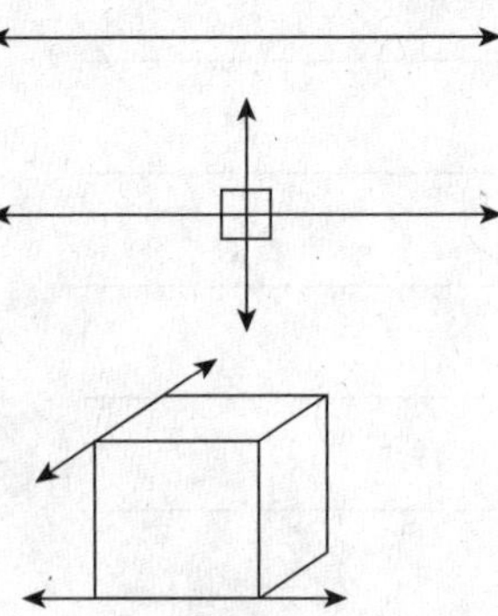

Use the figure to complete the statements.

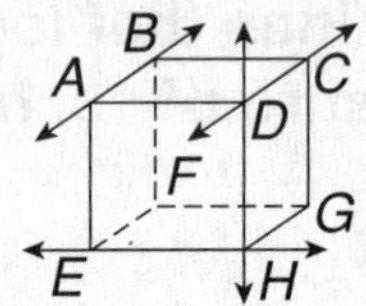

1. $\overleftrightarrow{AB}$ and $\overleftrightarrow{DC}$ are ________________ lines, since they never meet and lie in the same plane.

2. $\overleftrightarrow{DH}$ and $\overleftrightarrow{EH}$ are ________________ lines, since they intersect to form right angles.

3. $\overleftrightarrow{AB}$ and $\overleftrightarrow{EH}$ are ________________ lines, since they do not intersect and are not parallel.

A **transversal** is a line that intersects two or more lines.

- When two parallel lines are intersected by a transversal, 8 angles are formed.
- The 4 acute angles are congruent, and the 4 obtuse angles are congruent.
- The sum of an acute angle and an obtuse angle is 180°.

Use the figure to answer the questions.

4. Which angles are acute angles?

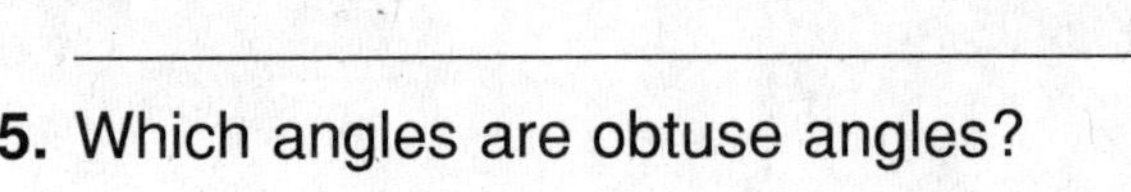

5. Which angles are obtuse angles?

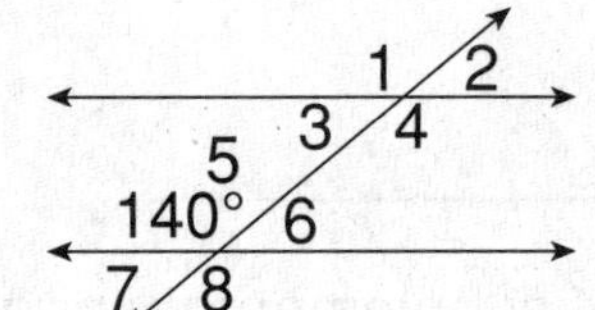

____

6. Find the measures of ∠1 and ∠2.

____

Name ______________________ Date __________ Class __________

LESSON 8-3

Challenge

Transversal Hints

In the diagram below, $\overline{KL}$ is parallel to $\overline{MN}$, and $\angle JKL$ is congruent to $\angle JLK$.

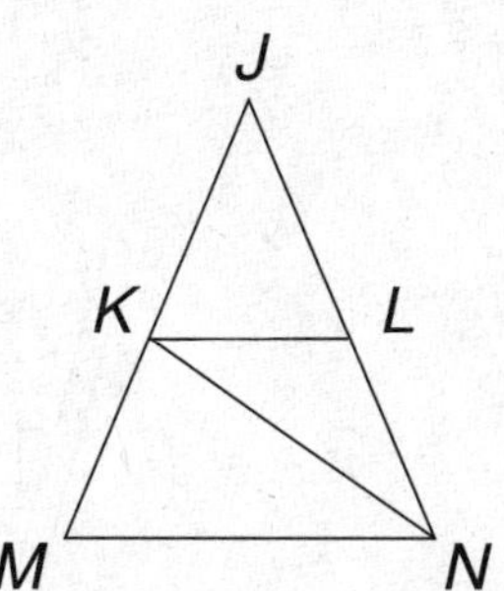

If $\angle JKL = 55°$, $\angle KNM = 33°$, and $\angle MJN = 70°$, find the following angle measures.

1. $\angle JMN$ __________

2. $\angle LKN$ __________

3. $\angle MKN$ __________

4. $\angle JLK$ __________

5. $\angle MNL$ __________

6. $\angle LNK$ __________

In the diagram below, line *A* is parallel to line *B*.

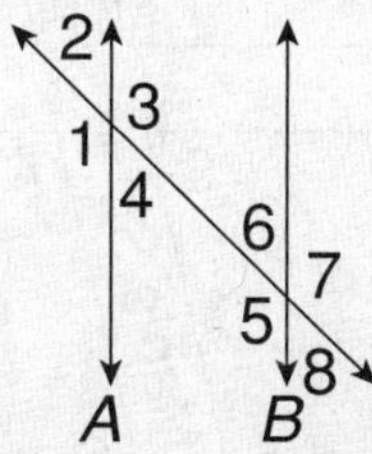

If $\angle 2 = x°$ and $\angle 7 = 3x°$, find the following angle measures.

7. $\angle 2$ __________

8. $\angle 7$ __________

9. $\angle 3$ __________

10. $\angle 5$ __________

11. $\angle 8$ __________

12. $\angle 6$ __________

Name ______________________ Date ____________ Class ____________

LESSON 8-3

Problem Solving

Angle Relationships

Write the correct answer.

In the drawing of the chair, the seat is parallel to the floor.

1. What is the measure of ∠1?

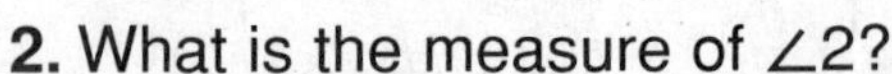

2. What is the measure of ∠2?

____

3. What is the measure of ∠3?

____

4. What is the measure of ∠4?

____

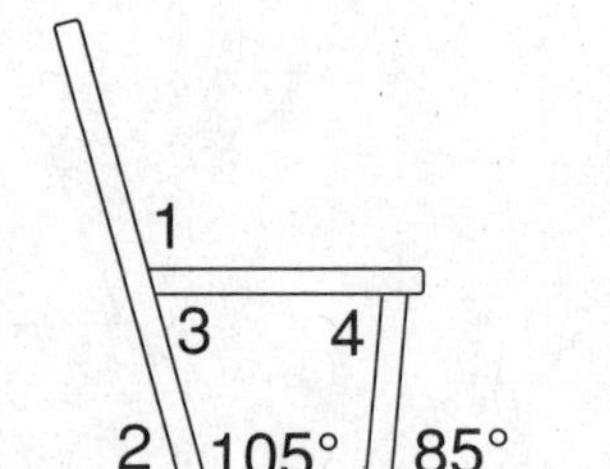

Choose the letter for the best answer.

The map shows the area around Falcon Park. Birch Street and Orchard Street are parallel to each other.

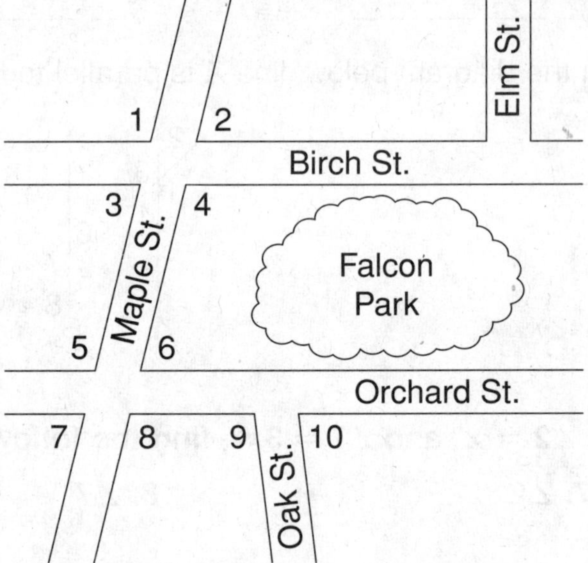

5. If ∠4 measures 112°, what is the measure of ∠6?

A 112° **C** 68°

B 22° **D** 108°

6. Which two angles are vertical angles?

F ∠2 and ∠3 **H** ∠2 and ∠4

G ∠2 and ∠6 **J** ∠2 and ∠5

7. If ∠10 measures 87°, what is the measure of ∠9?

A 77° **C** 87°

B 93° **D** 103°

8. Which is a transversal to Birch and Orchard streets?

F Maple Street **H** Oak Street

G Elm Street **J** Falcon Park

9. If ∠4 measures 112°, what is the measure of ∠1?

A 22° **C** 108°

B 68° **D** 112°

Name ______________________________ Date ____________ Class ____________

LESSON 8-3

Reading Strategies

Vocabulary Development

Lines that never meet and are the same distance apart are called **parallel lines.** The lines on sheet music are similar to parallel lines.

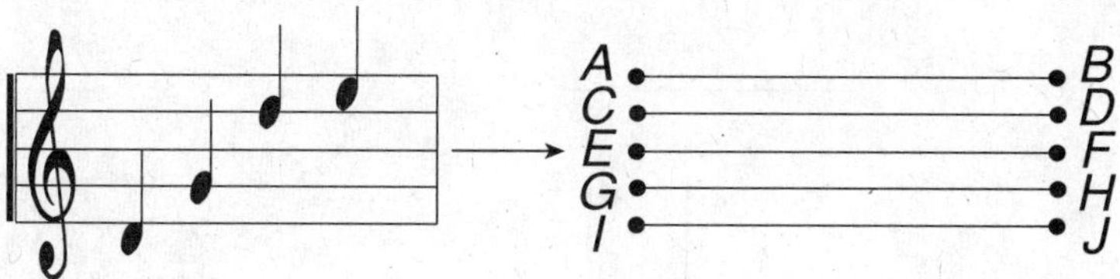

$\overline{EF} \parallel \overline{GH}$

Read: "Segment *EF* is parallel to segment *GH*."

1. Identify another pair of parallel line segments in the figure above.

____

2. How would you describe parallel lines?

____

____

Lines that meet to form square corners are called **perpendicular lines.**

Angles formed by these square corners measure 90°. $\overrightarrow{WY} \perp \overleftrightarrow{ZW}$

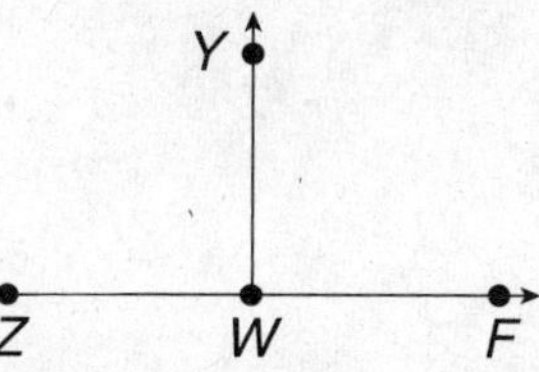

Read: "Ray *WY* is perpendicular to line *ZW*."

3. How can you tell if two lines, rays or segments are perpendicular?

____

4. Identify another pair of perpendicular lines or rays in the figure.

____

Lines that lie in different planes, but do not cross and are not parallel, are called **skew** lines.

$\overline{WC}$ and $\overline{TV}$ are skew.

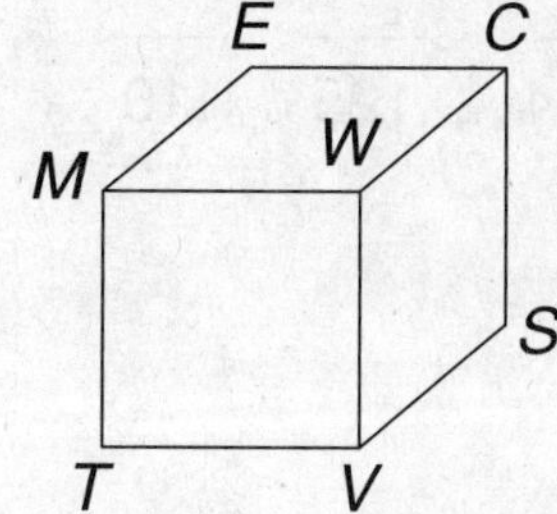

5. Identify two other line segments that are skew.

____

Name ________________________________ Date ____________ Class ____________

LESSON 8-3

Puzzles, Twisters & Teasers

A Little Birdie Told Me!

Find the measure of each angle with a question mark. Each has a corresponding letter. Match the letters to the answers to solve the riddle. Assume lines that look parallel are parallel.

T ____________

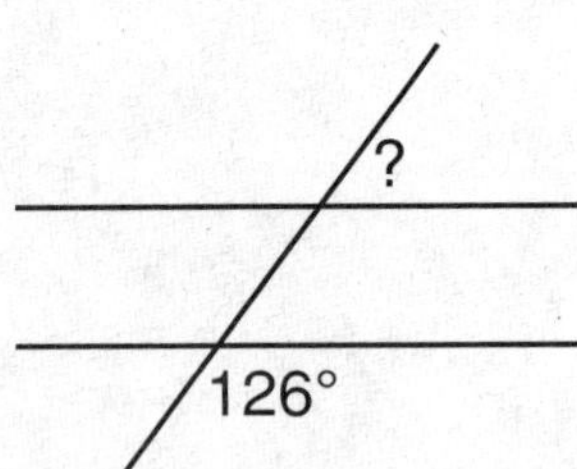

N ____________

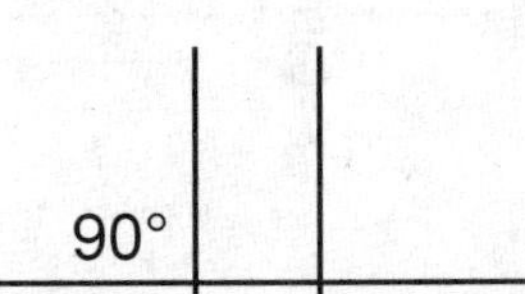

E ____________

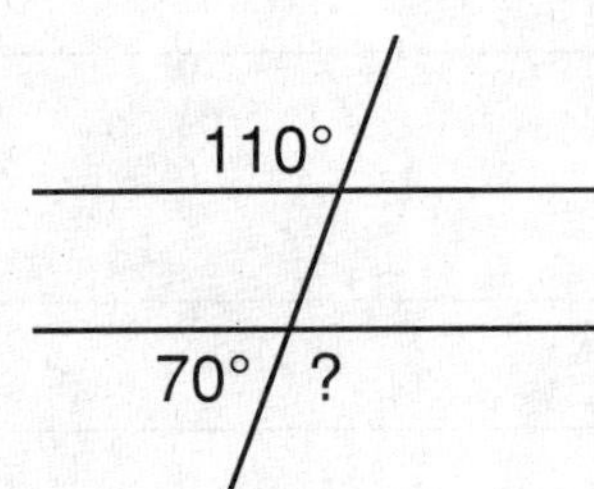

W ____________

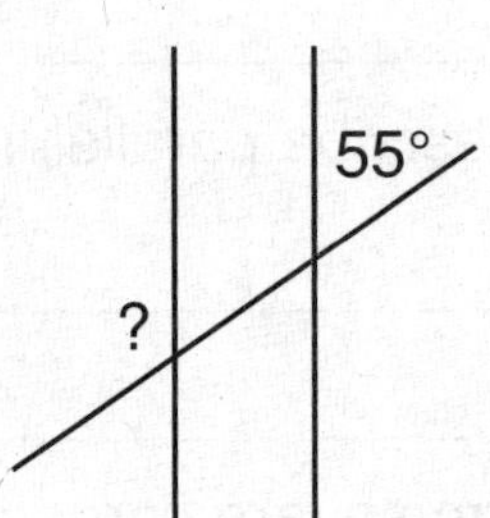

T ____________

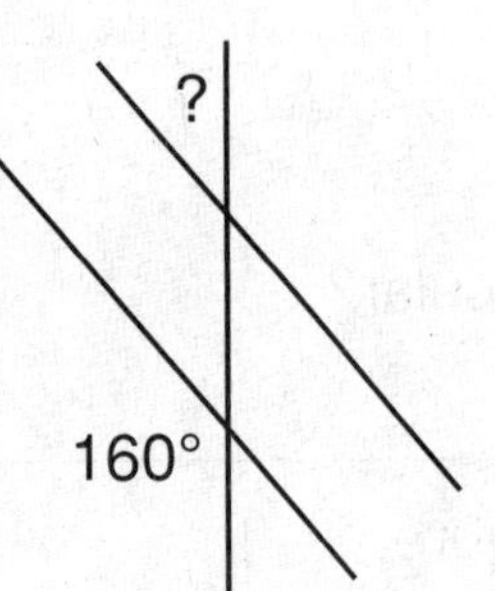

E ____________

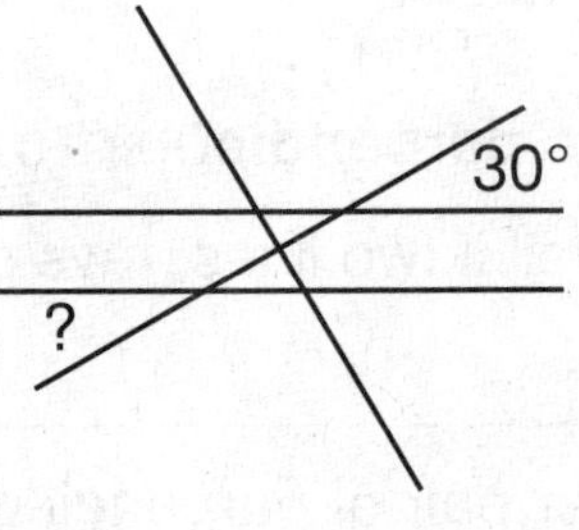

What do you give a sick bird?

A ___ ___ ___ **E** ___ **M** ___ ___ **T**

54 125 110 20 30 90

Name ______________________ Date ____________ Class ____________

LESSON 8-4

Practice A

Properties of Circles

Use circle C. Choose the letter for the best answer.

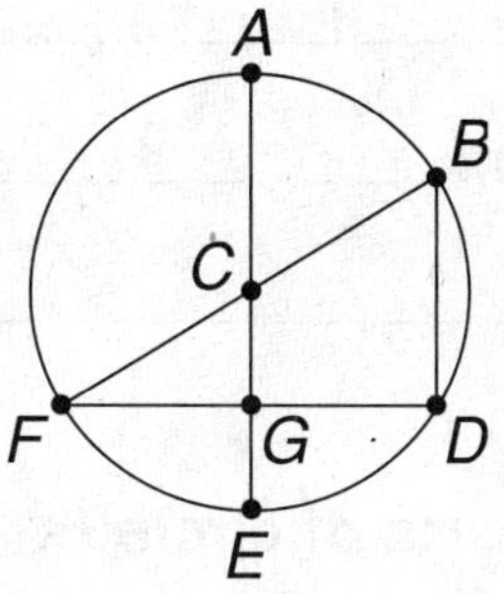

1. Which of the following is a radius?

A $\overline{BD}$ **C** $\overline{CE}$

B $\overline{FB}$ **D** $\overline{GE}$

2. Which of the following is a diameter?

F $\overline{FD}$ **H** $\overline{BC}$

G $\overline{AE}$ **J** $\overline{CG}$

3. Which of the following is a chord?

A $\overline{FD}$ **C** $\overline{BC}$

B $\overline{CA}$ **D** $\overline{GF}$

Name the parts of circle *K*.

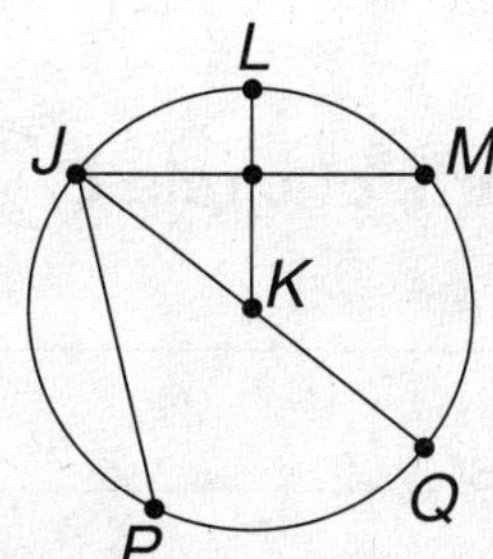

4. radii ______________________

5. diameters ______________________

6. chords ______________________

Name the parts of circle *T*.

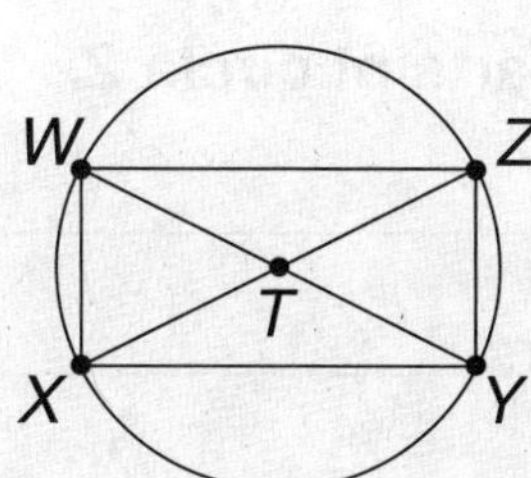

7. radii ______________________

8. diameters ______________________

9. chords ______________________

Use the circle graph.

10. The circle graph shows the uses of land in New Zealand. Find the central angle of the sector that shows the percent of land New Zealand uses for permanent pastures.

____

New Zealand Land Use

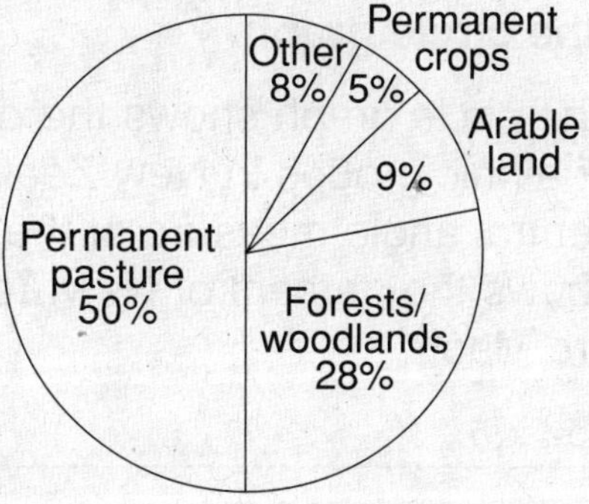

Name ______________________ Date ____________ Class ____________

LESSON 8-4

Practice B

Properties of Circles

Name the parts of circle *A*.

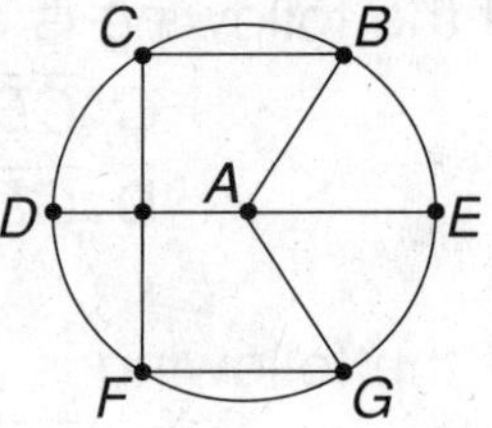

1. radii ______________________

2. diameters ______________________

3. chords ______________________

Name the parts of circle *H*.

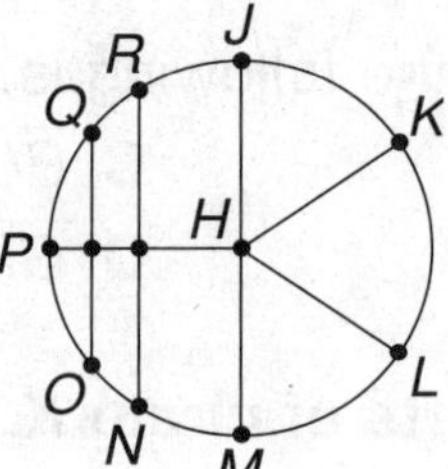

4. radii ______________________

5. diameters ______________________

6. chords ______________________

Name the parts of circle *C*.

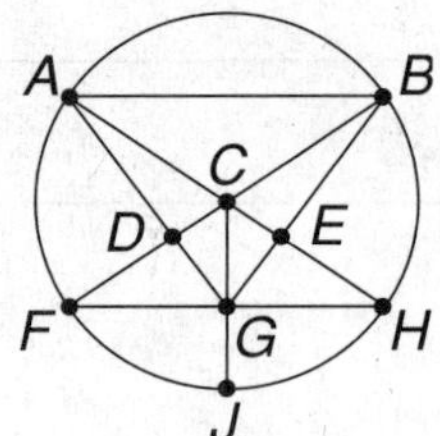

7. radii ______________________

8. diameters ______________________

9. chords ______________________

Name the parts of circle *Z*.

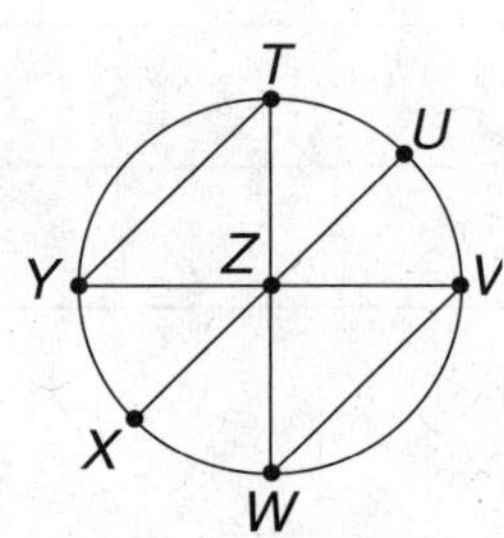

10. radii ______________________

11. diameters ______________________

12. chords ______________________

Use the circle graph.

13. The circle graph shows the distribution of ethnic groups in New Zealand. Find the central angle measure of the sector that shows the percent of New Zealanders who are Maori.

____

Name ______________________ Date ____________ Class ____________

LESSON 8-4

Practice C

Properties of Circles

Use the circle graph.

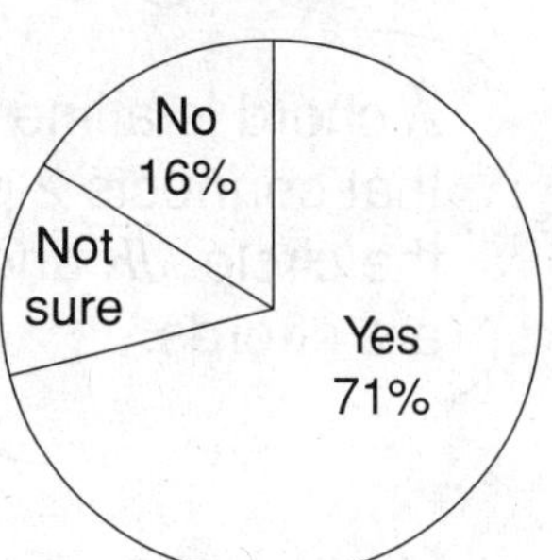

1. An Internet poll asked people if they believe that taking vitamins will make them live longer. The circle graph shows the response. Find the central angle of the sector that shows the percent of people who answered "No."

____

2. Find the central angle of the sector that shows the percent of people who answered "Yes."

____

3. Find the central angle of the sector that shows the percent of people who answered "Not sure."

____

Name the parts of circle *N*.

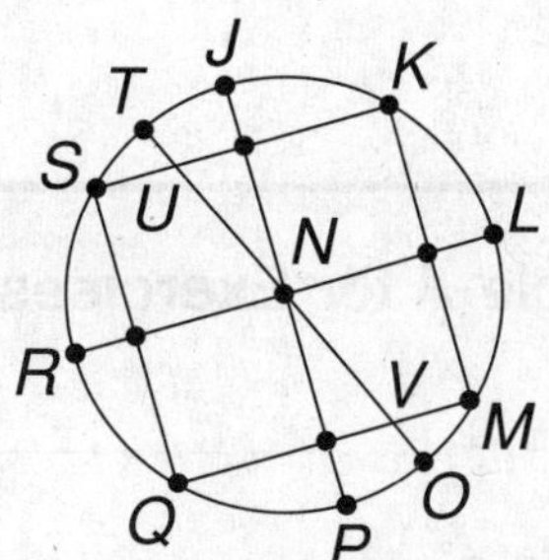

4. radii _______________________________

5. diameters ___________________________

6. chords ______________________________

7. If $\overline{SK} \parallel \overline{QM}$ and $m\angle NUK = 58°$, what is the measure of $\angle NVQ$?

____

8. If $m\angle NUK = 58°$, what is the measure of $\angle SUN$?

____

Use the clock for Exercises 9 and 10.

9. Find the central angle measure between the minute and hour hands on the clock.

____

10. Find the central angle measure between the second and hour hands on the clock.

____

Name ______________________ Date ____________ Class ____________

LESSON 8-4

Reteach

Properties of Circles

A chord is a line segment that connects 2 points on the circle. $\overline{JK}$ and $\overline{GH}$ are chords.

A diameter is a chord that passes through the center of the circle. $\overline{PR}$ is a diameter.

A circle is named by its center. This is circle *O*.

A radius is a line segment that connects a point on the circle with the center of the circle. $\overline{AO}$ and $\overline{BO}$ are radii.

Use circle *A* for Exercises 1–4.

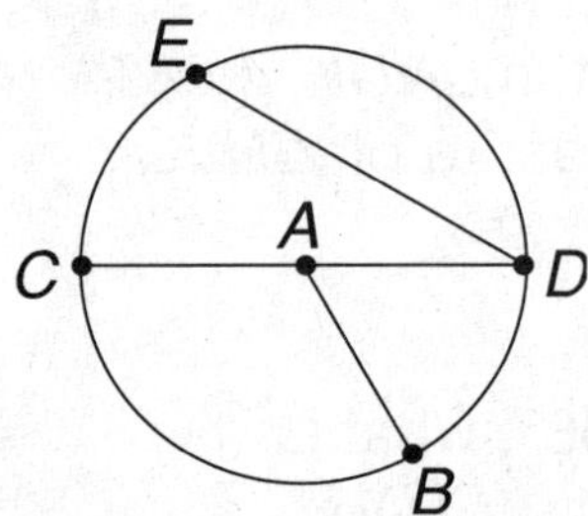

1. $\overline{AB}$ is a ____________________.

2. $\overline{CD}$ is a ____________________.

3. $\overline{DE}$ is a ____________________.

4. $\overline{CA}$ is a ____________________.

Use circle *Z* for Exercises 5–7.

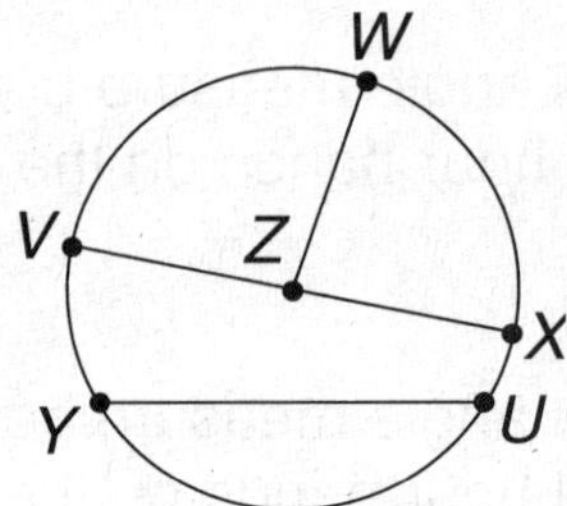

5. Name 3 radii of circle *Z*.

____

6. Name 2 chords of circle *Z*.

____

7. Name the diameter of circle *Z*.

____

Name ________________________________ Date ____________ Class ____________

LESSON 8-4

Challenge

Arc Length

An *arc* is a portion of the circumference of a circle. On the circle below, arc *QR* corresponds to the part of the circumference whose endpoints are the radii forming central angle *QCR*.

A *minor arc* is less than 180° and is named by its endpoints. Arc *QR* is a minor arc. A *major arc* is greater than 180° and is named by its endpoints and one other point that lies on the arc. Arc *QSR* is a major arc.

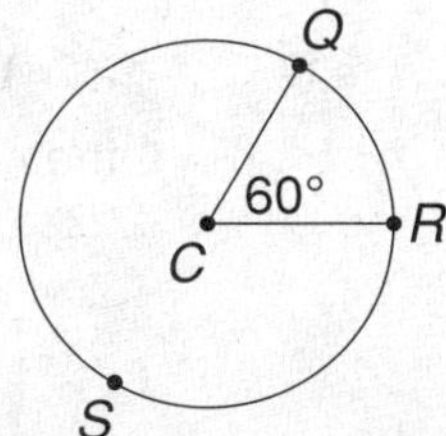

The degree measure of a minor arc is the measure of its central angle. The degree measure of a major arc is 360° minus the degree measure of the minor arc.

For circle *C* above, the measure of arc *QR* is 60°, and the measure of arc *QSR* is 360° − 60°, or 300°.

Find the degree measure of each arc.

1.

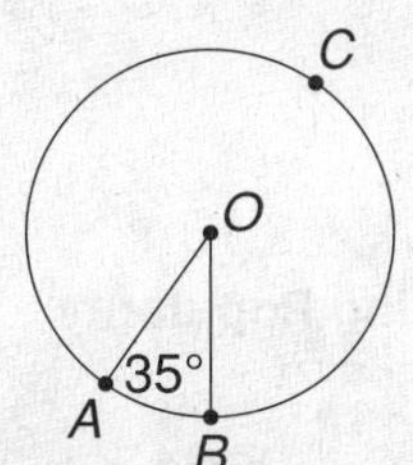

arc *BCA* ____________

2.

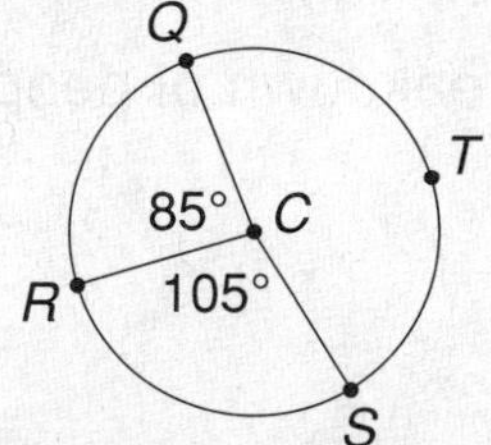

arc *QRS* ____________

3.

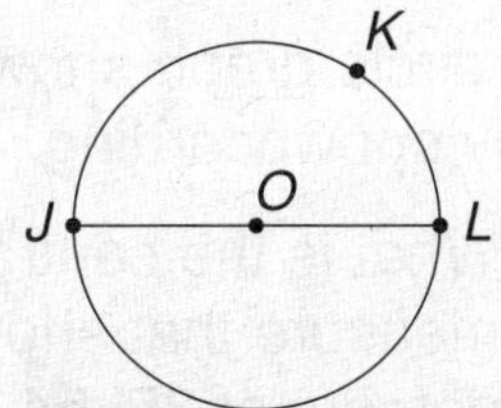

arc *JKL* ____________

4.

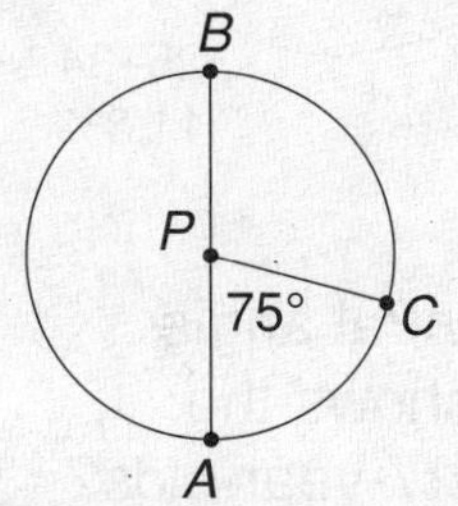

arc *BC* ____________

5.

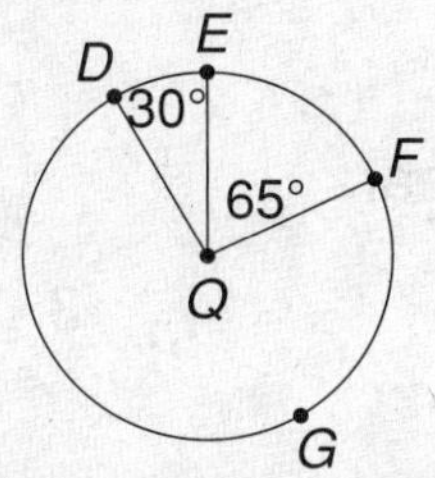

arc *DGF* ____________

6.

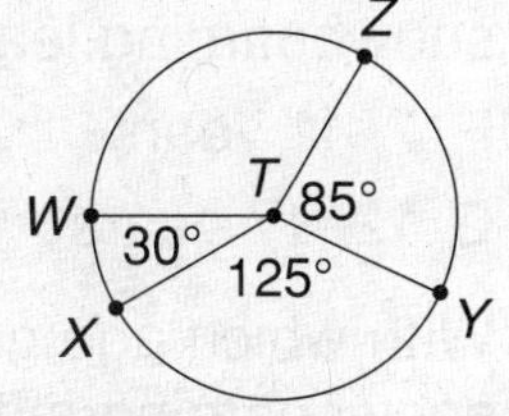

arc *WZ* ____________

Name ______________________ Date ____________ Class ____________

LESSON 8-4

Problem Solving

Properties of Circles

Write the correct answer.

The circle graph shows the results of a survey in which people were asked to name their hobbies.

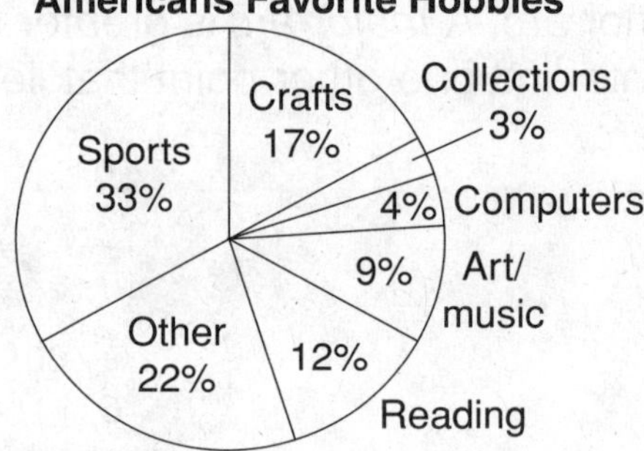

1. If 1,000 people were surveyed, how many said that collecting things is their favorite hobby?

____

2. Find the central angle measure of the sector that shows the percent of people who named sports as their favorite hobby.

____

3. Find the central angle measure of the sector that shows the percent of people who named reading as their favorite hobby.

____

4. Find the central angle measure of the sector that shows the percent of people who like to do crafts as their favorite hobby.

____

Choose the letter for the best answer.

The circle graph shows the age breakdown of people who most enjoy snowboarding.

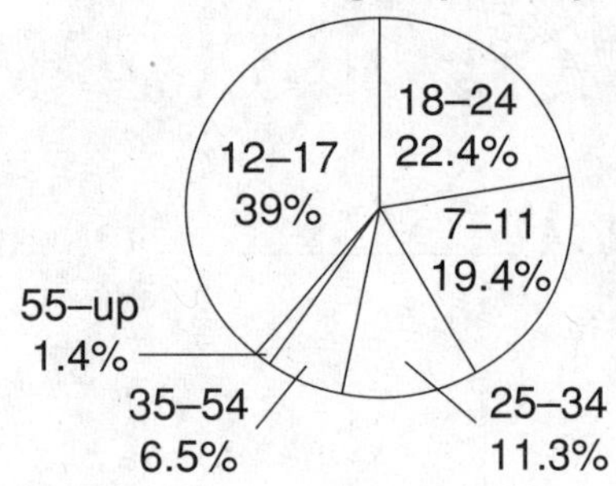

5. What is the central angle measure that shows the percent of 35–54-year-olds?

A 5.1° **C** 23.4°

B 6.5° **D** 40.7°

6. With which age group is snowboarding least popular?

F 7–11 years **H** 25–34 years

G 12–17 years **J** 55–up

7. With which age group is snowboarding most popular?

A 7–11 years **C** 18–24 years

B 12–17 years **D** 25–34 years

8. What is the central angle measure that shows the percent of 12–17-year-olds?

F 39° **H** 108°

G 140.4° **J** 180°

Name ______________________________ Date ___________ Class ___________

LESSON 8-4

Reading Strategies

Focus On Vocabulary

A **circle** is a closed figure. All points on the circle are the same distance from the **center point.** The center point is called the center of the circle.

The **diameter** is a line segment that passes through the center of a circle. It has its endpoints on the circle.

The **radius** is a line segment from the center of a circle to a point on the circle.

The **chord** is a line segment with endpoints on the circle.

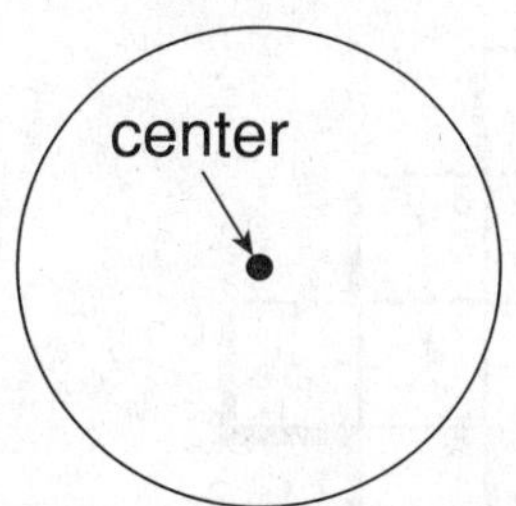

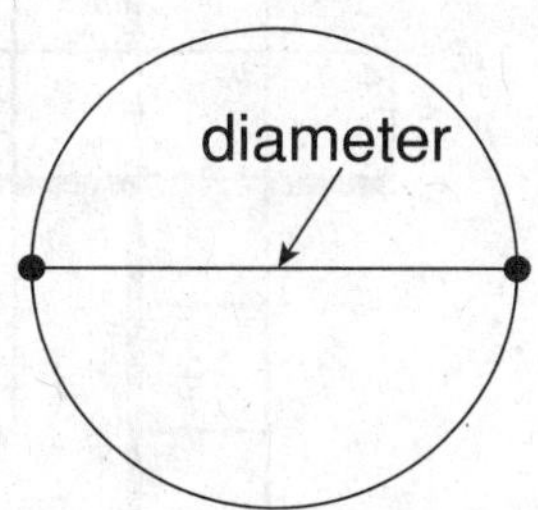

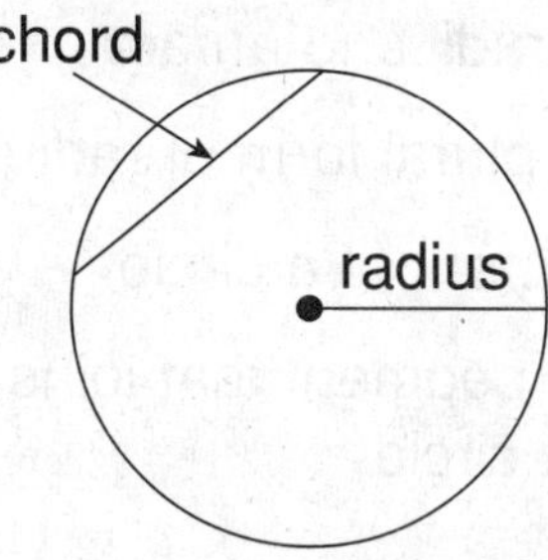

Answer each question.

1. What name is given to the line segment that passes through the center of a circle and has endpoints on the circle?

____

2. What is the center point of a circle called?

____

3. What name is given to a line segment with endpoints on the circle?

____

4. Is a circle an open figure or a closed figure?

____

5. What is the name of the line segment that extends from the center of the circle to a point on the circle?

____

Name ______________________ Date __________ Class __________

LESSON 8-4

Puzzles, Twisters & Teasers

Going in Circles

Across

1. In a plane, the set of all points that are the same distance from a given point
4. The part of a circle enclosed by two radii and an arc
5. The plural form of radius
7. Segment of a circle
8. Line segment that joins two points on a circle

Down

2. The point from which all points on a circle are equidistant
3. Angle formed by two radii
6. The straight line which passes through the center of a circle and has endpoints on the circle

Name ______________________ Date __________ Class __________

LESSON 8-5

Practice A

Classifying Polygons

Determine whether each figure is a polygon. If it is not, explain why not.

1. 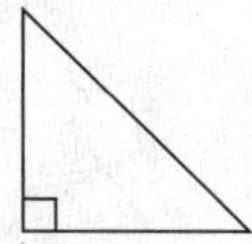

____

2. 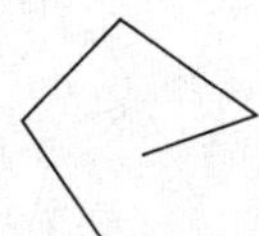

____

3. 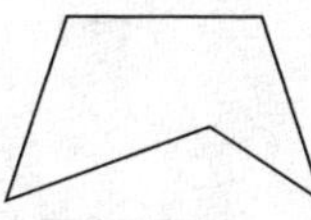

____

4. 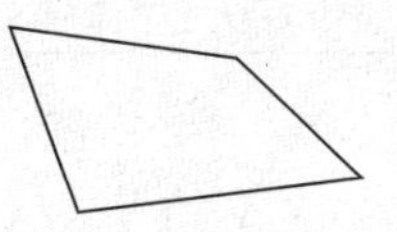

____

5. 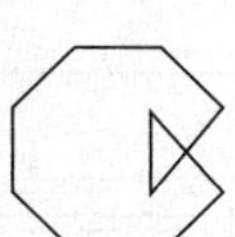

____

6. 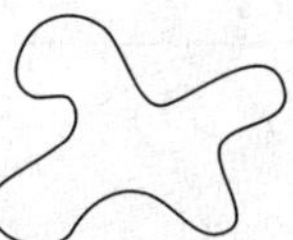

____

Draw a line from each polygon to its correct name.

7. 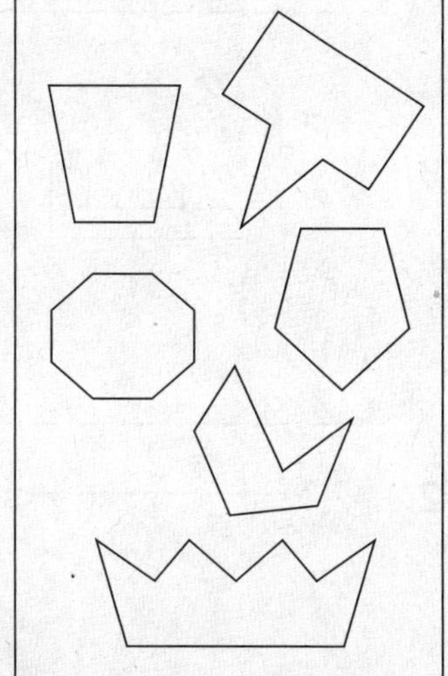

octagon

quadrilateral

heptagon

pentagon

nonagon

hexagon

Determine whether each figure is a regular polygon. If it is not, explain why not.

8. 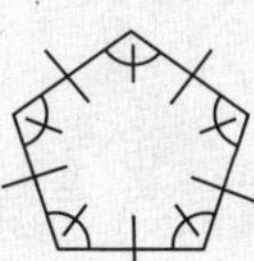

____

____

9. 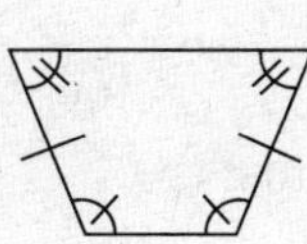

____

____

10. 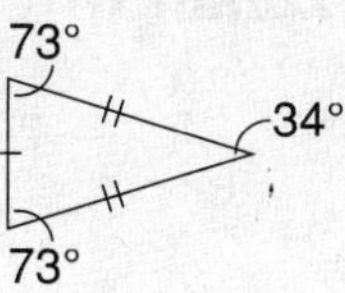

____

____

Name ______________________ Date __________ Class __________

LESSON **8-5**

Practice B

Classifying Polygons

Determine whether each figure is a polygon. If it is not, explain why not.

1.

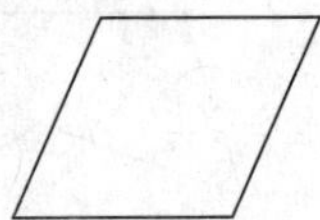

2.

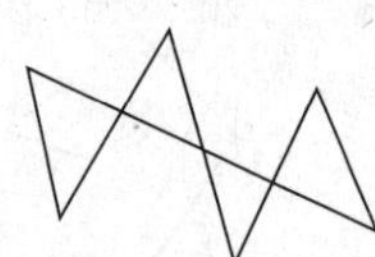

3.

4.

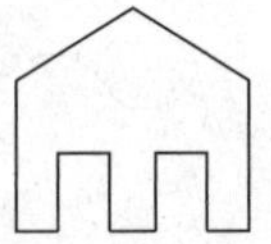

5.

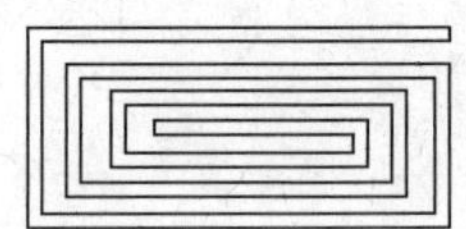

6.

Name each polygon.

7.

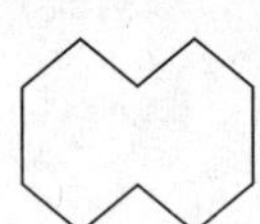

8.

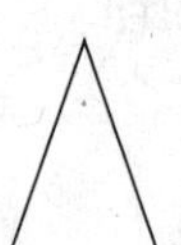

9.

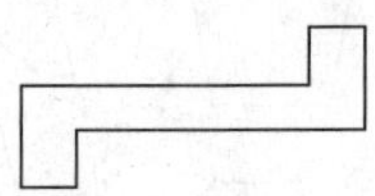

10.

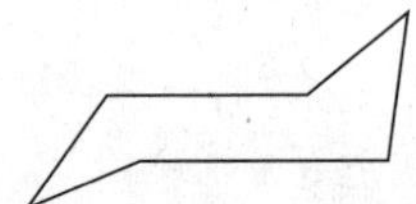

11.

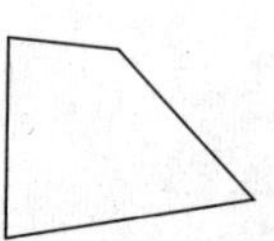

12.

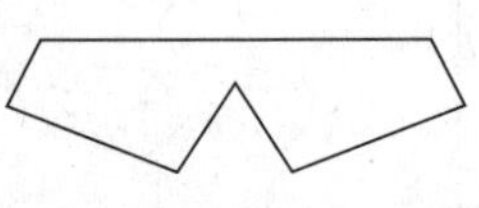

Name each figure and tell whether it is a regular polygon. If it is not, explain why not.

13.

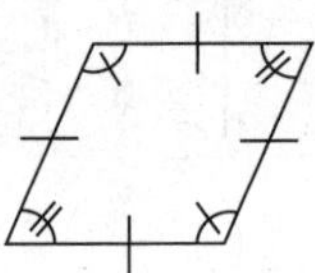

14.

108°
108° 108°
108° 108°

15.

45° 6 ft 45°
6 ft 6 ft
135° 135°
14 ft

Name ______________________ Date ___________ Class ___________

LESSON 8-5

Practice C
Classifying Polygons

Determine whether each figure is a polygon. If it is not, explain why not.

1. 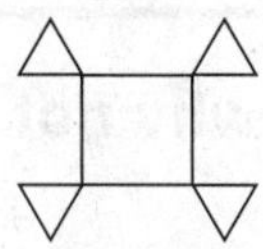
____

2. 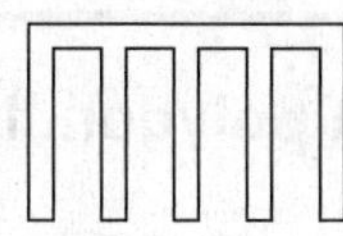
____

3. 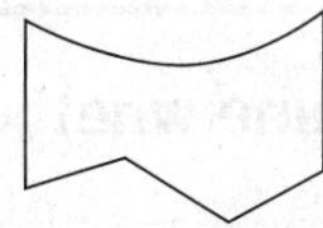
____

4. 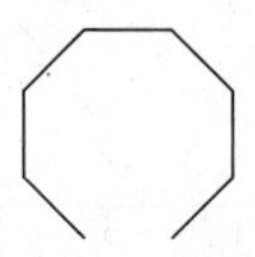
____

5. 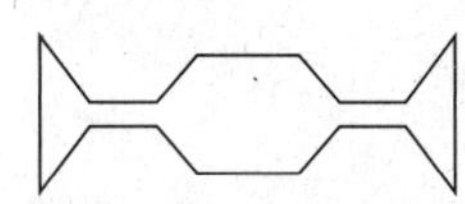
____

6. 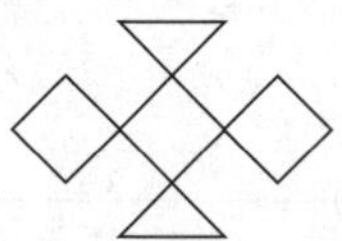
____

Name each polygon.

7. 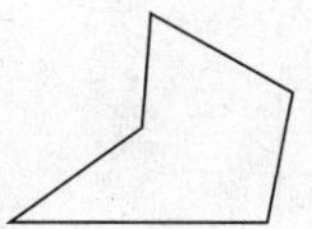
____

8. 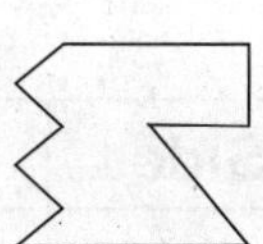
____

9. 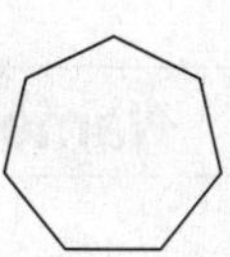
____

10. 
____

11. 
____

12. 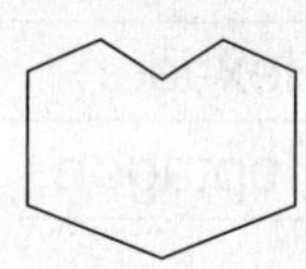
____

Determine if the figures in each exercise can be put together to form a regular polygon. If so, name the regular polygon.

13. 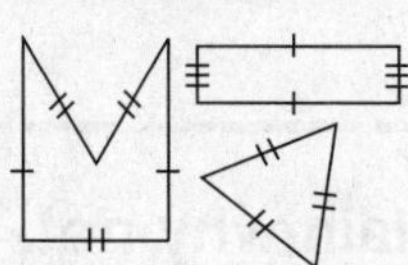
____

14. 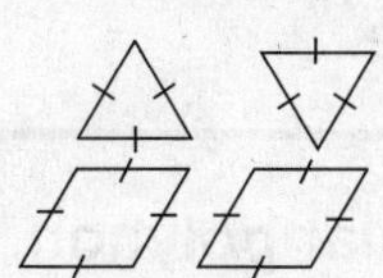
____

15. 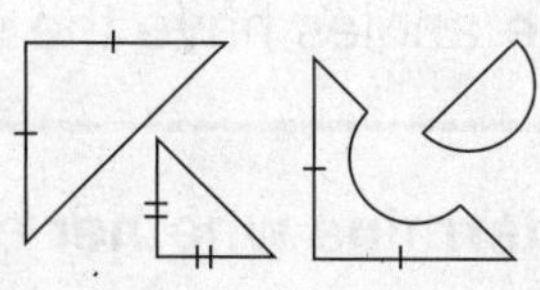
____

Name ______________________ Date __________ Class __________

LESSON 8-5

Reteach
Classifying Polygons

A **polygon** is a closed figure that starts and stops at the same point. A polygon is made up of line segments that do not cross.

Determine whether each figure is a polygon. If it is not, explain why not.

1. 

2. 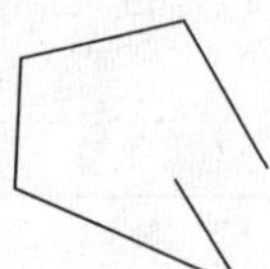

3. 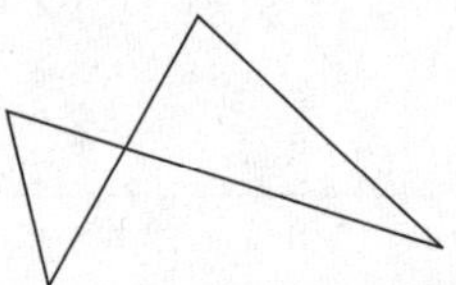

Polygons can be classified by the number of sides or angles. The number of sides and angles is the same.

Complete the table.

	Name	Number of Sides	Number of Angles
4.		3	3
5.	Quadrilateral		4
6.	Pentagon		5
7.	Hexagon	6	
8.	Heptagon		7
9.		8	8
10.	Nonagon	9	
11.	Decagon		10

All the sides of a **regular polygon** have the same length, and all the angles have the same measure.

Determine whether each is a regular polygon. If it is not, explain why not.

12. 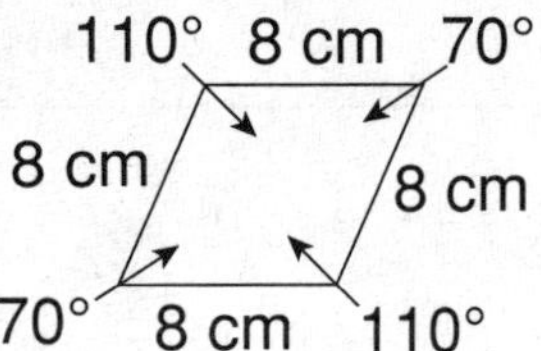

13.

120° 2 in. 120°
2 in. 2 in.
120° 120°
2 in. 2 in.
120° 2 in. 120°

14.

9 in.
4 in.
90° 90°
90° 90°

Name ______________________ Date __________ Class __________

LESSON 8-5

Challenge

The Diagonal Count

Consecutive vertices connect the sides of a polygon. A *diagonal* of a polygon is a segment that connects two nonconsecutive vertices.

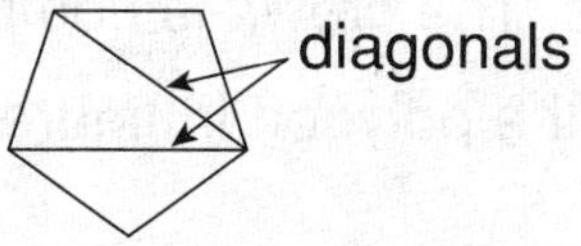

The figure shows two diagonals of a pentagon.

diagonals

Other diagonals can be drawn in the pentagon from other vertices.

For each polygon below, find the total number of diagonals from all the vertices. Do not count the same diagonal more than once. *Hint:* Draw each figure and then draw all possible diagonals. Use different colors to distinguish the diagonals as the number of sides of the polygons increases.

1. Triangle ______________

2. Quadrilateral ______________

3. Pentagon ______________

4. Hexagon ______________

5. Heptagon ______________

6. Octagon ______________

Name ______________________ Date __________ Class __________

LESSON 8-5

Problem Solving

Classifying Polygons

Write the correct answer.

The drawing shows a crown designed by a child in an arts and crafts class. The crown is composed of 4 different figures.

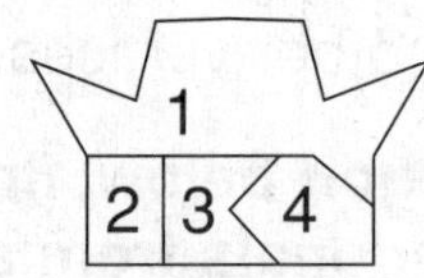

1. Name the polygon in figure 1.

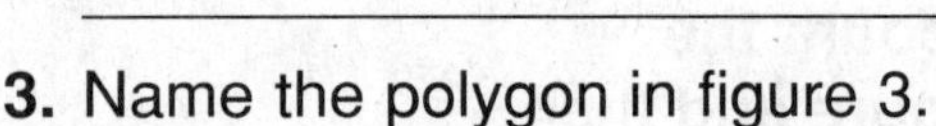

2. Name the polygon in figure 4.

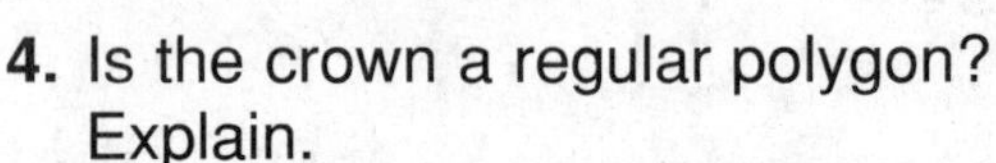

3. Name the polygon in figure 3.

____

4. Is the crown a regular polygon? Explain.

____

____

5. Name the polygon formed by figures 2, 3, and 4.

____

____

Choose the letter for the best answer.

The box shows some basic shapes from a word processing tool bar.

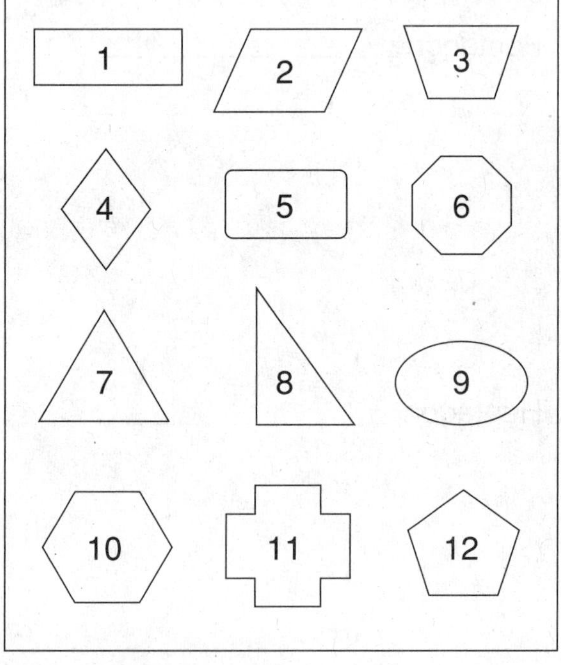

6. Which figure is *not* a quadrilateral?
   A Figure 2 **C** Figure 4
   B Figure 3 **D** Figure 5

7. Which figure is *not* a polygon?
   F Figure 3 **H** Figure 11
   G Figure 9 **J** Figure 12

8. Which figure is a pentagon?
   A Figure 4 **C** Figure 10
   B Figure 5 **D** Figure 12

9. Which figure is a regular polygon?
   F Figure 1 **H** Figure 7
   G Figure 2 **J** Figure 8

10. Which figure is an octagon?
    A Figure 6 **C** Figure 11
    B Figure 10 **D** Figure 12

Name ______________________ Date __________ Class __________

LESSON 8-5

Reading Strategies

Compare and Contrast

A **polygon** is a closed, plane figure formed by three or more line segments.

A **regular polygon** has sides of equal length, and all angle measures are the same.

Compare the regular and irregular polygons.

Regular Polygon **Irregular Polygon**

Answer each question.

1. How many sides do both shapes have? __________
2. How many angles do both shapes have? __________
3. How are the two polygons alike?

____

4. How are the two polygons different?

____

____

Compare this regular octagon and regular pentagon.

Regular Pentagon Regular Octagon

5. Compare the number of sides and angles in a pentagon to the number of sides and angles in an octagon.

____

____

6. How are the octagon and pentagon alike?

____

____

7. How are the octagon and pentagon different?

____

Name ______________________________ Date ____________ Class ____________

LESSON 8-5

Puzzles, Twisters & Teasers

Bean Me Up, Scotty!

Decide whether or not each figure is a polygon. Circle the letters next to your answers. Unscramble the letters to solve the riddle.

1.

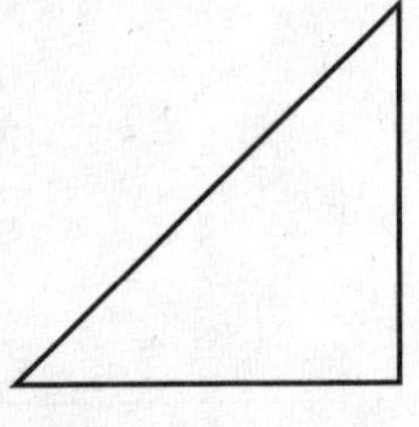

polygon **L** not polygon **P**

2.

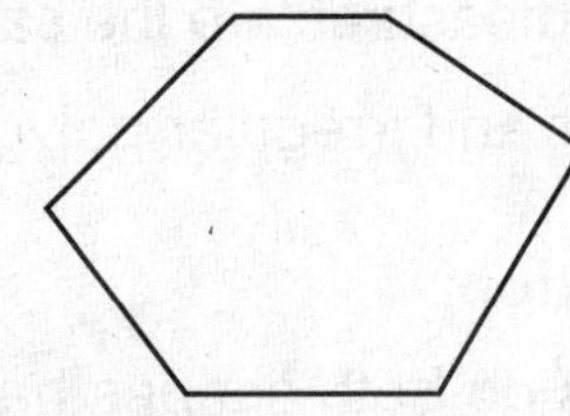

polygon **A** not polygon **D**

3.

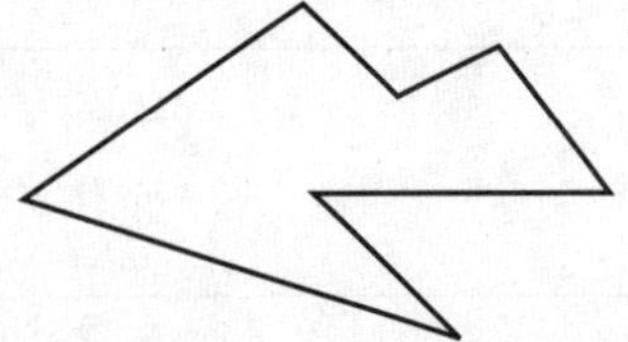

polygon **M** not polygon **C**

4.

polygon **P** not polygon **E**

5.

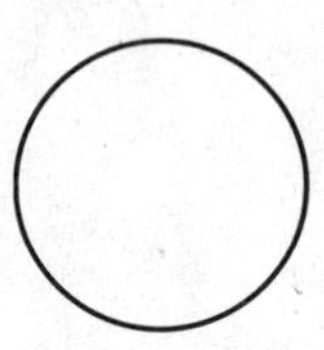

polygon **T** not polygon **A**

6.

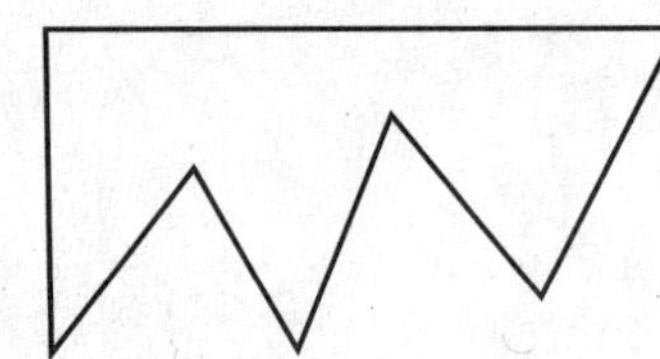

polygon **N** not polygon **E**

What type of beans do llamas eat?

___ **L** ___ ___ **A** **B** ___ ___ ___ **S**

Name ________________________ Date ____________ Class ____________

LESSON **8-6**

Practice A

Classifying Triangles

Classify each triangle according to its sides and angles. Choose the terms needed from the box.

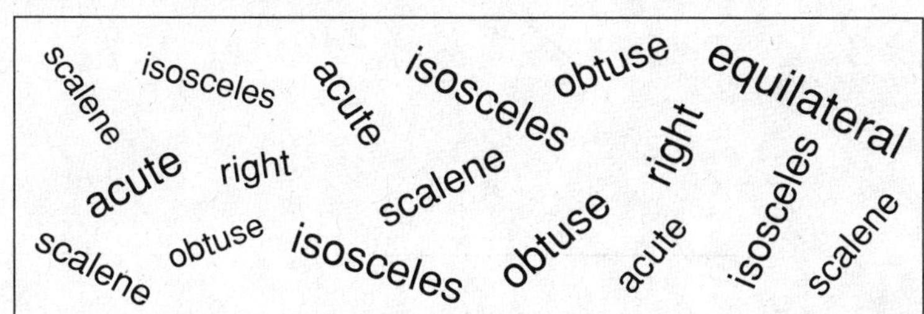

1.

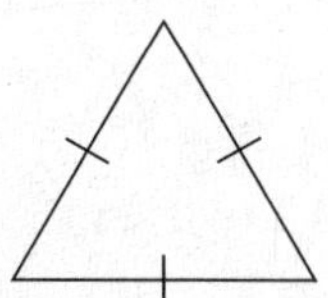

____

2.

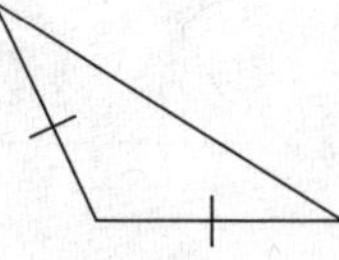

____

3.

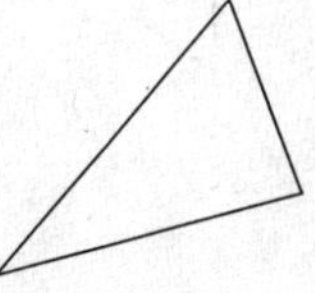

____

4.

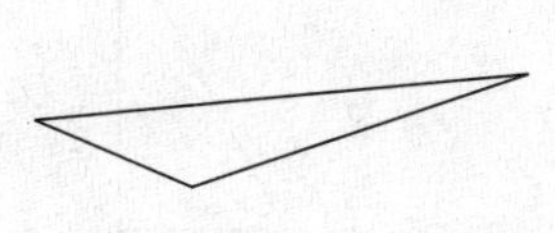

____

5.

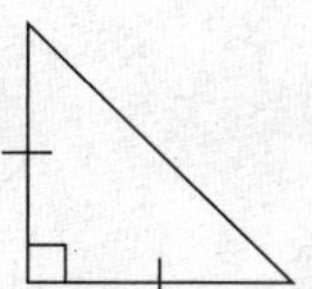

____

6.

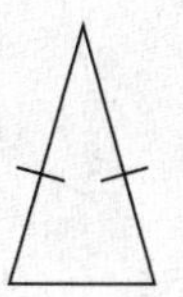

____

7.

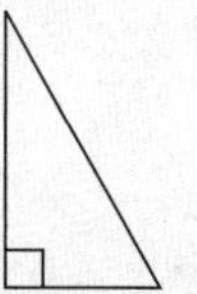

____

8.

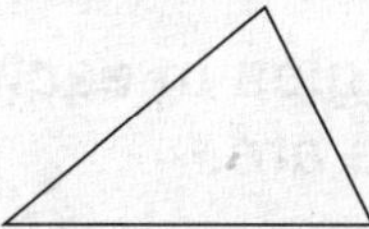

____

9.

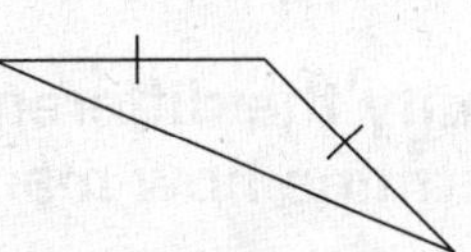

____

Describe the different kinds of triangles in each figure by their sides and angles and determine how many of each there are.

10.

____

____

____

11.

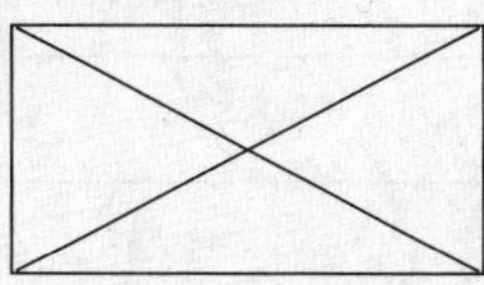

____

____

____

Name ______________________ Date ____________ Class ____________

LESSON 8-6

Practice B

Classifying Triangles

Classify each triangle according to its sides and angles.

1. 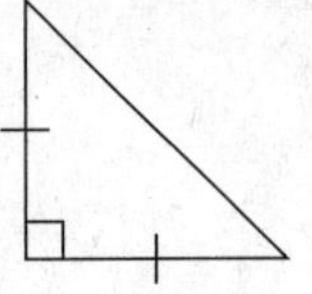

____

2. 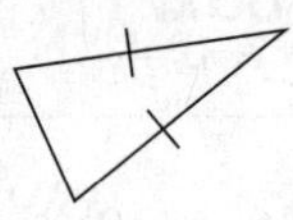

____

3. 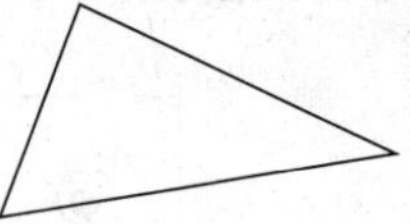

____

4. 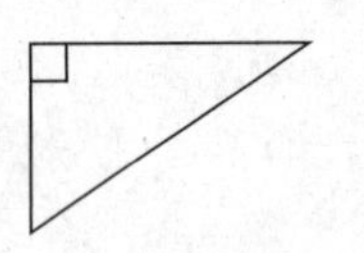

____

5. 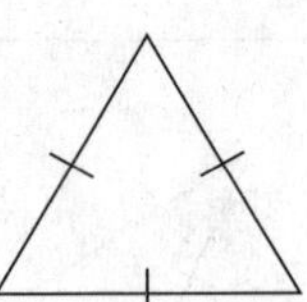

____

6. 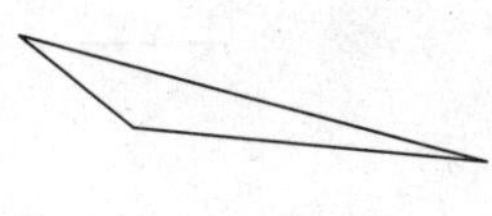

____

7. 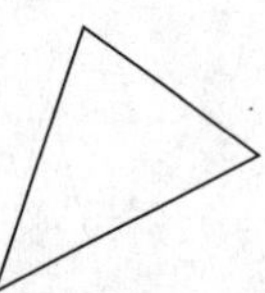

____

8. 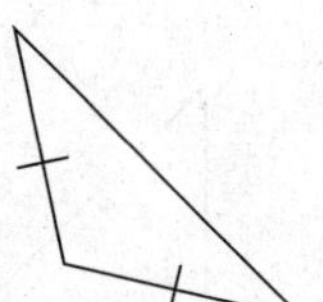

____

9. 

____

Identify the different types of triangles in each figure and determine how many of each there are.

10. 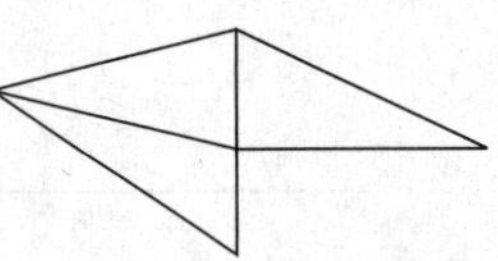

____________________ ________

____

____

____

11. 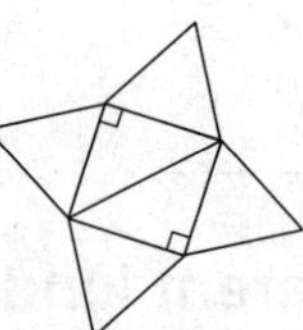

____________________ ________

____

____

____

Name ______________________ Date __________ Class __________

LESSON 8-6

Practice C

Classifying Triangles

Classify each triangle according to its sides and angles.

1.

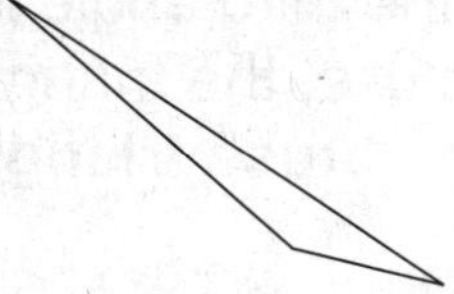

____

2.

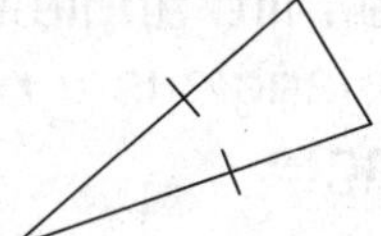

____

3.

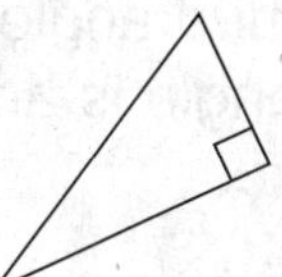

____

Classify each triangle according to the lengths of its sides.

4. 9.9 cm, 9 cm, 0.99 cm ______________

5. 40 ft, 4 ft, 40 ft ______________

6. 22 ft, 22 ft, 21 ft ______________

7. 2.3 in., 2.3 in., 2.3 in. ______________

8. 5.6 m, 5.06 m, 5.16 m ______________

9. 19 cm, 18 cm, 19 cm ______________

Classify each triangle according to the measures of its angles.

10. 25°, 75°, 80° ______________

11. 178°, 1°, 1° ______________

12. 90°, 40°, 50° ______________

13. 45°, 40°, 95° ______________

14. 23°, 90°, 67° ______________

15. 55°, 60°, 65° ______________

Identify the different types of triangles in each figure and determine how many of each there are.

16.

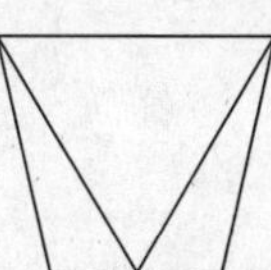

____

____

17.

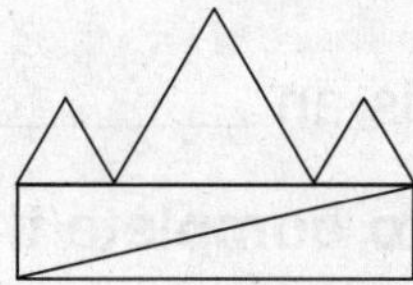

____

____

Name ______________________ Date ____________ Class ____________

LESSON **8-6**

Reteach

Classifying Triangles

All triangles have at least two acute angles.

If the third angle is acute, the triangle is an **acute triangle.**

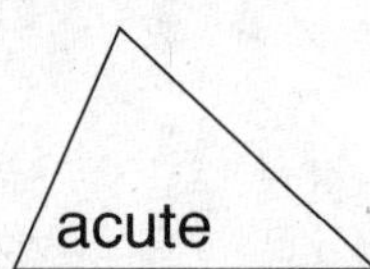

If the third angle is right, the triangle is a **right triangle.**

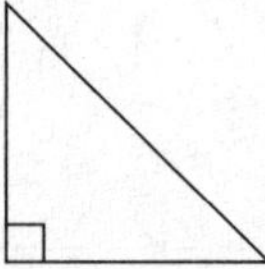

If the third angle is obtuse, the triangle is an **obtuse triangle.**

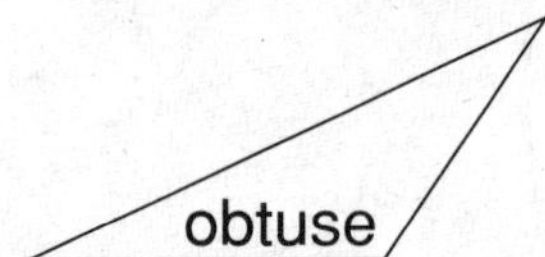

A triangle is **scalene** if each side is a different length.

4 in.
5 in.
6 in.

A triangle is **isosceles** if two sides are the same length.

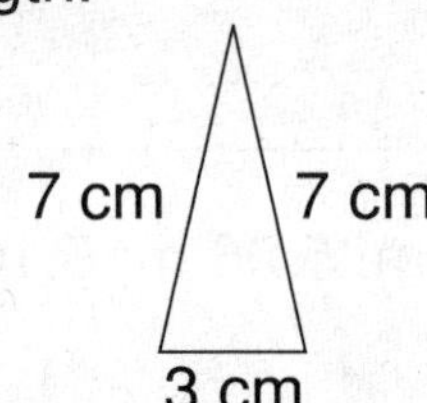

A triangle is **equilateral** if all three sides are the same length.

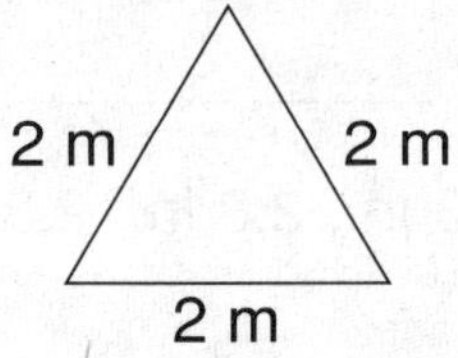

Use the figure to complete the statements.

5 ft
60°
5 ft
60° 60°
5 ft

1. Each of the 3 angles is less than 90°, so the triangle is an ______________ triangle.
2. The triangle has 3 sides that have the same length, so it is an ______________ triangle.

Use the figure to complete the statements.

45°
10 cm
14 cm
90° 45°
10 cm

3. The triangle has a 90° angle, so it is a ______________ triangle.
4. The triangle has 2 sides that have the same length, so it is an ______________ triangle.

Use the figure to complete the statements.

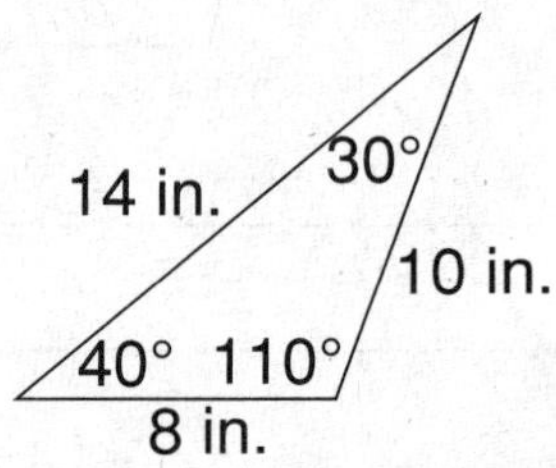

5. The triangle has an angle that measures more than 90°, so it is an ______________ triangle.
6. Each side of the triangle has a different length, so it is a ______________ triangle.

Name ______________________ Date __________ Class __________

LESSON 8-6

Challenge

Finding Shapes

1. How many triangles can you find?

____

2. How many triangles can you find?

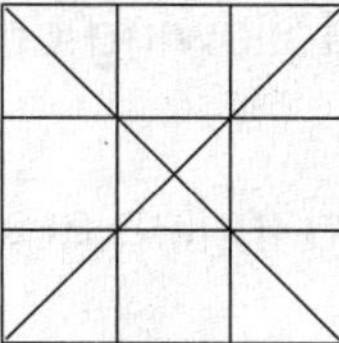

____

3. How many quadrilaterals can you find?

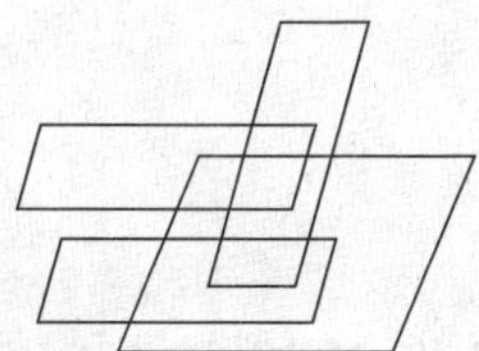

____

____

4. How many triangles and quadrilaterals can you find?

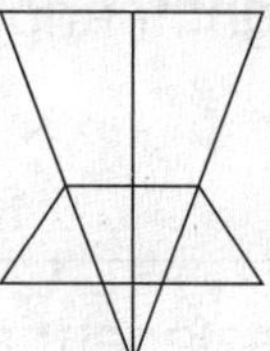

____

____

Name a shape in the diagram as described.

5. a right isosceles triangle

____

6. a quadrilateral containing 2 right angles

____

7. a right scalene triangle

____

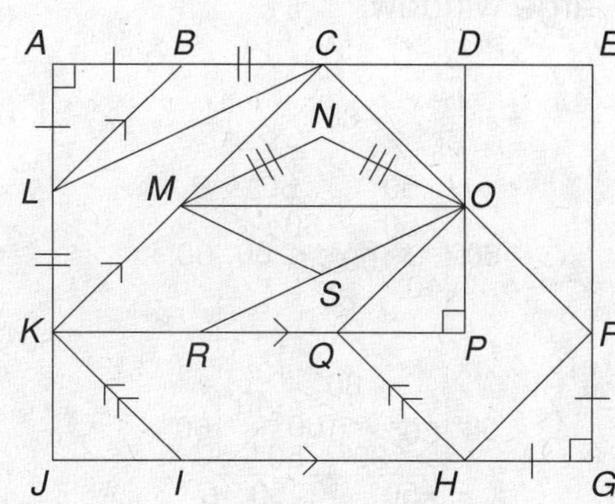

8. quadrilateral containing an obtuse angle and no right angles

____

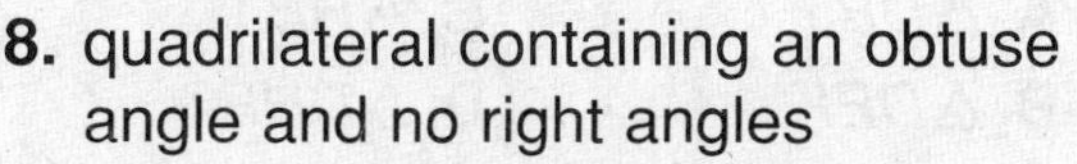

9. a quadrilateral with 2 pairs of equal sides

____

10. an obtuse triangle

____

11. an isosceles triangle that is not a right triangle

____

Name ______________________ Date __________ Class __________

LESSON 8-6

Problem Solving

Classifying Triangles

Write the correct answer.

Brian made the following drawing of his kite.

1. How many triangles are in the figure?

____

2. How many acute triangles are in the figure?

____

3. How many right triangles are in the figure?

____

4. How many scalene triangles are in the figure?

____

5. How many isosceles triangles are in the figure?

____

Choose the letter for the best answer.

The figure shows an architect's design for a large window.

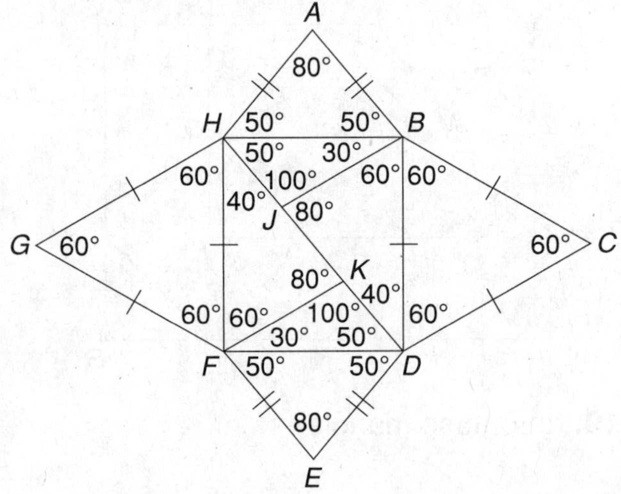

6. Which triangle is an equilateral triangle?

   A △*ABH** **C** △*FGH*

   B △*BDH* **D** △*DHF*

7. Which triangle is a right triangle?

   F △*FHK* **H** △*DKF*

   G △*DEF* **J** △*DBH*

8. Which triangle is an isosceles triangle?

   A △*DBJ* **C** △*HBJ*

   B △*DEF* **D** △*DHF*

9. Which triangle is a scalene triangle?

   F △*ABH* **H** △*BCD*

   G △*HBJ* **J** △*DEF*

10. Which triangle is an obtuse triangle?

    A △*DKF* **C** △*JBD*

    B △*HKF* **D** △*ABH*

Name ______________________ Date ____________ Class ____________

LESSON **8-6**

Reading Strategies

Make A Drawing

This chart will help you compare triangles by angles and by sides.

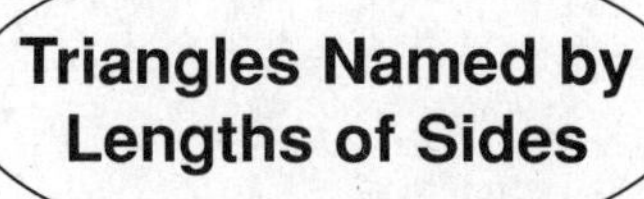

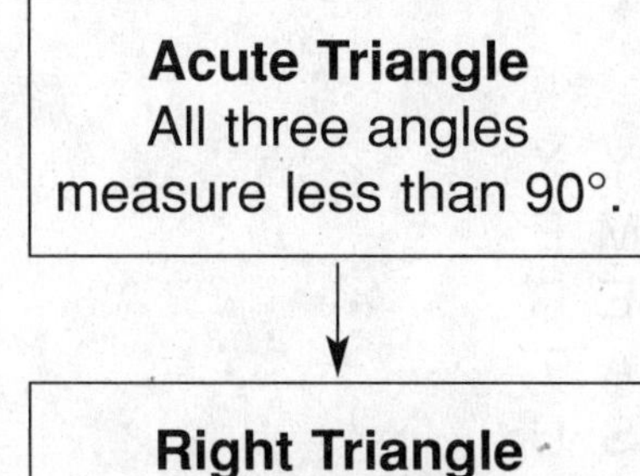

Triangles Named by Lengths of Sides	**Triangles Named by Angle Measures**
Equilateral Triangle All three sides are equal.	**Acute Triangle** All three angles measure less than 90°.
↓	↓
Isosceles Triangle Two sides are equal.	**Right Triangle** One angle measures 90°.
↓	↓
Scalene Triangle None of the sides are equal.	**Obtuse Triangle** One angle measures more than 90°.

This drawing shows:

- One angle measures 90° → right triangle
- Two sides are equal. → isosceles triangle

The drawing shows an isosceles right triangle.

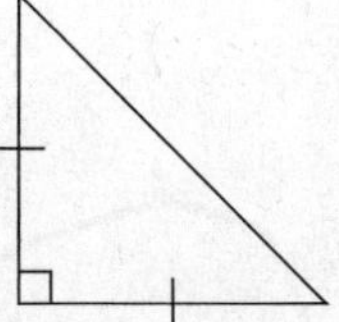

Draw the following triangles in the space provided.

1. A scalene obtuse triangle
2. An equilateral triangle

Name ______________________ Date __________ Class __________

LESSON **8-6**

Puzzles, Twisters & Teasers

You Are Correct, Sir!

Find and circle words from the word list in the word search (horizontally, vertically or diagonally). Find a word that answers the riddle. Circle it and write it on the line.

triangle	scalene	isosceles	equilateral	acute
obtuse	right	side	angle	measure

```
I N C O R R E C T L Y X
S O B T U S E X R C M E
O W E R T Y U L I E E R
S C A L E N E Q A M A T
C A C T G V U I N H S U
E C R I G H T O G F U P
L U W E R A N G L E R M
E T C V B S I D E B E Q
S E Q U I L A T E R A L
```

What is the only word in the dictionary that is spelled incorrectly?

____

Name ______________________ Date __________ Class __________

LESSON 8-7

Practice A

Classifying Quadrilaterals

1. List the five major special quadrilaterals.

____

____

Give all of the names that apply to each quadrilateral.

2.

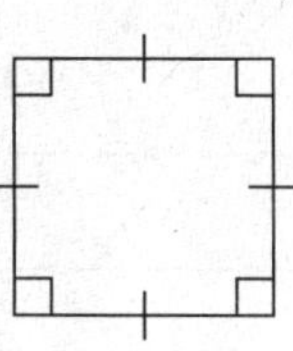

3.

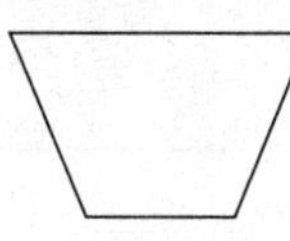

4.

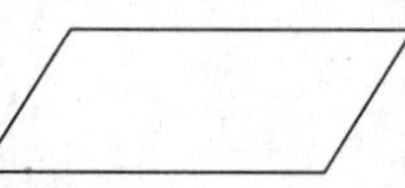

Give the name that best describes each quadrilateral.

5.

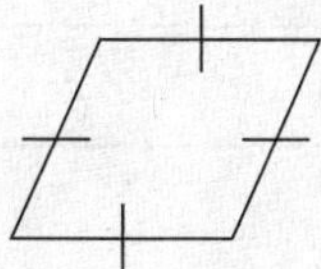

6.

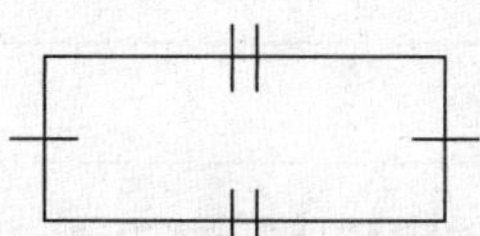

7.

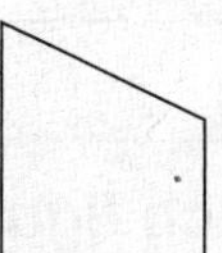

Draw each figure. If it is not possible to draw, explain why.

8. A rhombus that is also a square.

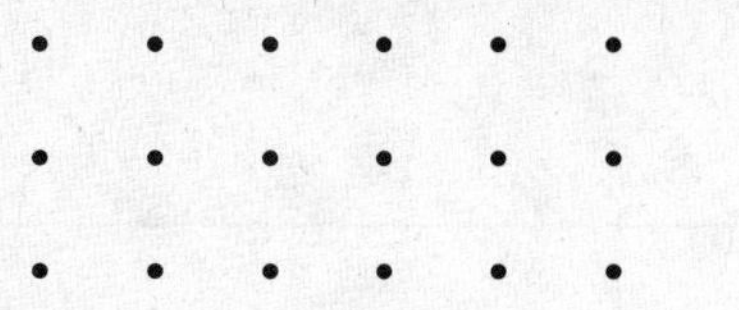

9. A trapezoid that is also a rectangle.

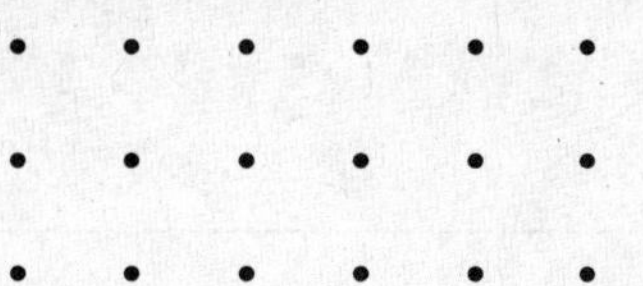

Name ______________________ Date __________ Class __________

LESSON 8-7

Practice B

Classifying Quadrilaterals

Give all of the names that apply to each quadrilateral. Then give the name that best describes it.

1. 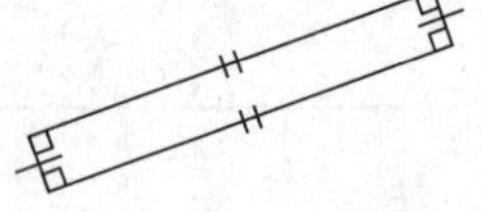

2. 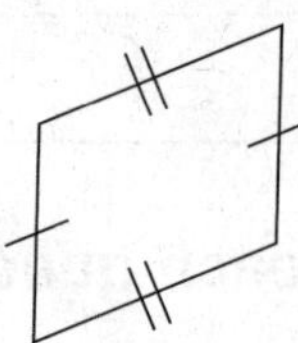

3. 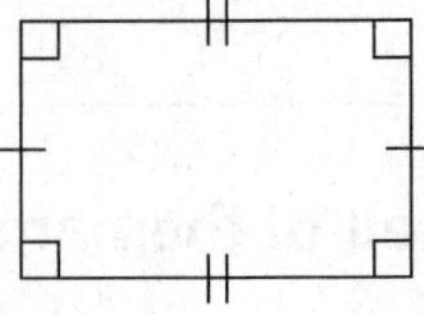

4. 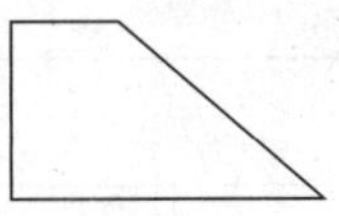

5. 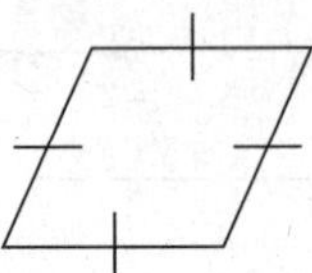

6. 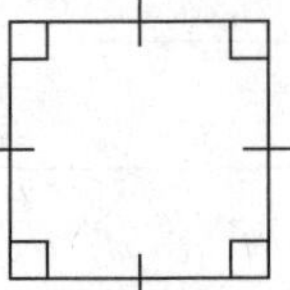

Draw each figure. If it is not possible to draw, explain why.

7. A rectangle that is not a parallelogram.

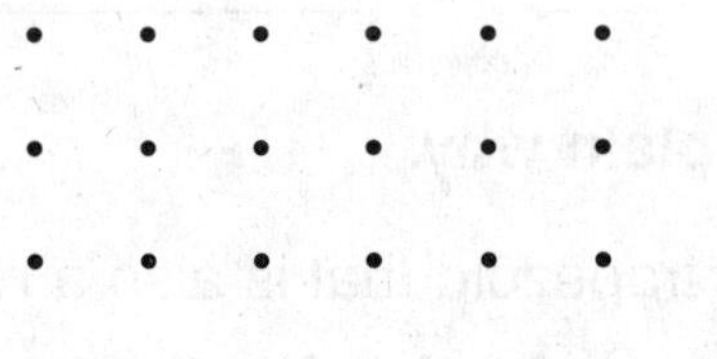

8. A rectangle that is not a square.

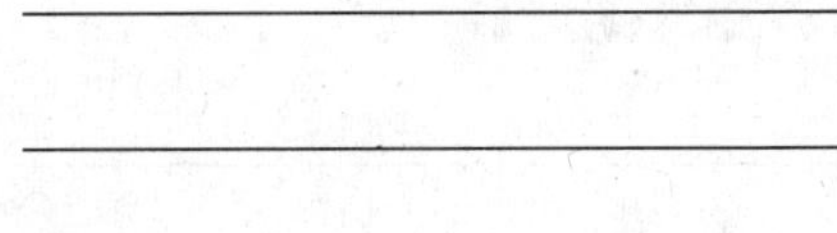

Name ______________________ Date ____________ Class ____________

LESSON 8-7

Practice C

Classifying Quadrilaterals

Give all of the names that apply to each quadrilateral. Then give the name that best describes it.

1.

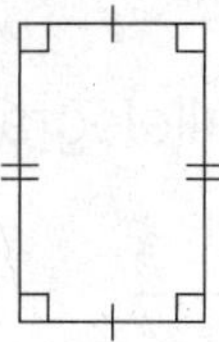

2.

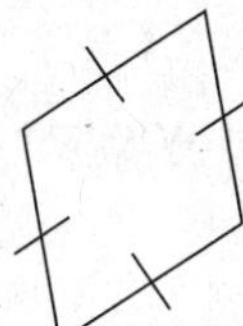

3.

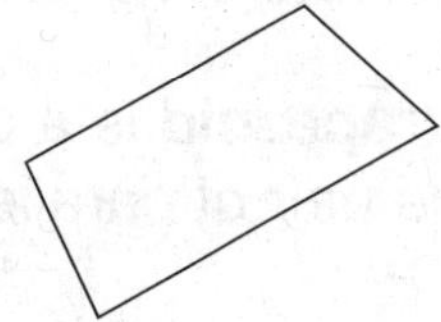

Draw each figure. If it is not possible to draw, explain why.

4. a rhombus that is not a rectangle

5. a trapezoid that is also a square

Name the types of quadrilaterals with each property.

6. exactly one pair of parallel sides ____________

7. four congruent sides ____________

8. Graph the points $A(3, 3)$, $B(-1, 5)$, $C(-5, -1)$, and $D(1, -4)$ and draw quadrilateral $ABCD$. What quadrilateral did you draw?

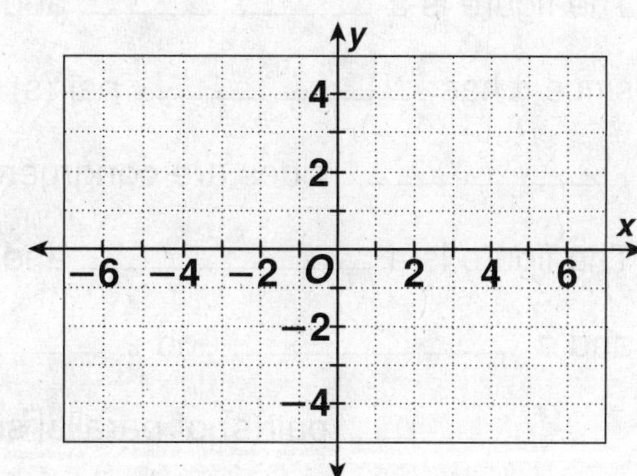

9. Graph the points $W(1, 5)$, $X(3, -2)$, $Y(1, -5)$, and $Z(-1, 2)$ and draw quadrilateral $WXYZ$. What quadrilateral did you draw?

Name ______________________ Date ____________ Class ____________

LESSON 8-7

Reteach

Classifying Quadrilaterals

Quadrilaterals are polygons that have 4 sides. In the figures below, arrows are used to indicate parallel sides, and tick marks are used to indicate congruent sides.

- A **trapezoid** is a quadrilateral with only one pair of parallel sides.

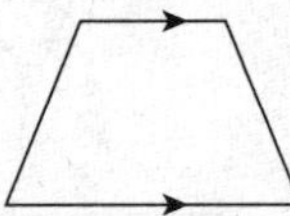

- A **parallelogram** is a quadrilateral with two pairs of parallel sides.

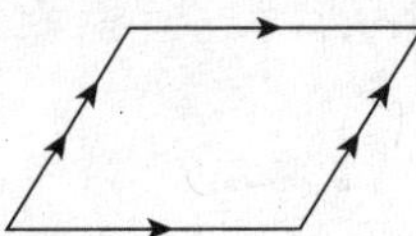

- A **rectangle** is a parallelogram with all four angles congruent.

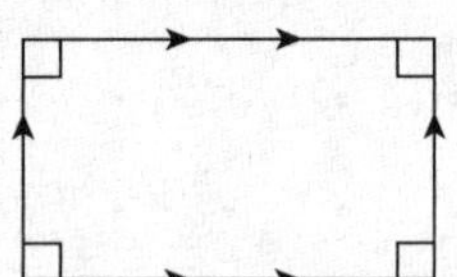

- A **rhombus** is a parallelogram with all four sides congruent.

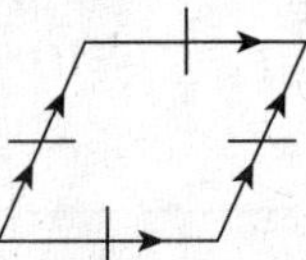

- A **square** is a rectangle with all four sides congruent.

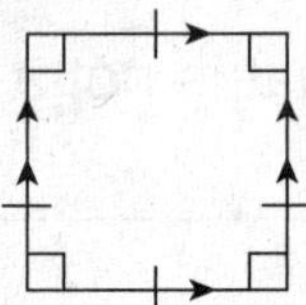

Use the figures to complete the statements.

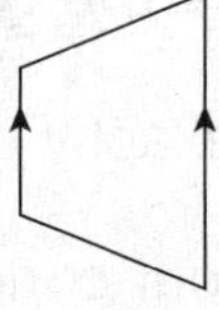

1. The figure is a ______________ since it has ______________ pair of parallel sides.

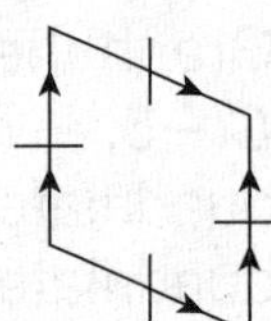

2. The figure is a ______________ and a ______________ since it has ______________ pair(s) of parallel sides, and ______________ sides are congruent.

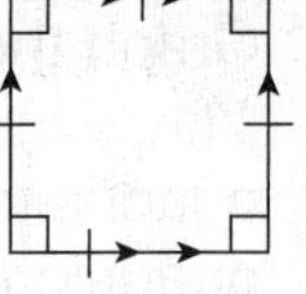

3. The figure is a ______________ and a ______________ and a ______________ and a ______________ since it has ______________ pair(s) of parallel sides, ______________ angles are congruent, and ______________ sides are congruent.

Name ______________________ Date ____________ Class ____________

LESSON 8-7

Challenge

It's Only a Quadrilateral!

Match each picture with the list showing which quadrilaterals are found in the picture.

1.

List A:
rhombus
trapezoid
rectangle

2.

List B:
trapezoid
rectangle
square
parallelogram

3.

List C:
rhombus
rectangle
square
parallelogram

4. Draw a design on the rug using only the following quadrilaterals: trapezoid, rectangle, and rhombus. You can use as many of each figure as you like, but all must be used at least once.

Name ______________________ Date __________ Class __________

LESSON 8-7

Problem Solving

Classifying Quadrilaterals

Write the correct answer.

1. A garden has 2 pairs of congruent sides. The congruent sides are not adjacent, and the angles are not all congruent. What is the shape of the garden?

____

2. A poster advertising an art show has 1 pair of parallel sides. All the angles are different, and none of the sides are congruent. What is the shape of this poster?

____

3. In a math report, Ming said that since a rhombus has 4 equal sides and a square has 4 equal sides, all rhombuses are squares. Is she correct? Explain.

____

____

4. A student was asked to give an example of a polygon that is not a quadrilateral and an example of a quadrilateral that is not a polygon. Is this possible? Why or why not?

____

____

Choose the letter for the best answer.

This is a drawing of Piet Mondrian's painting *Composition 8*.

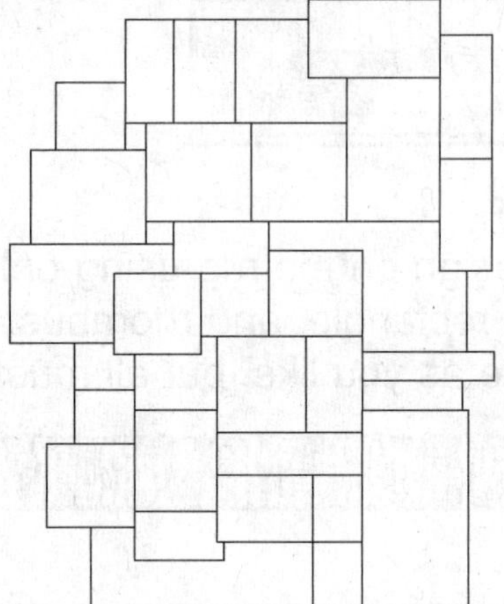

5. Which type of quadrilateral do you see most in the painting?

A rhombus **C** rectangle
B trapezoid **D** square

6. Which type of quadrilateral does not appear to be in the painting?

F rectangle
G parallelogram
H square
J trapezoid

7. Which statement is true about the painting?

A There are no parallelograms.
B There are only rectangles.
C There are only quadrilaterals.
D None of the above

8. Which statement is true about the painting?

F There are no triangles.
G There are only squares.
H There are no squares.
J None of the above

Name ______________________ Date __________ Class __________

LESSON 8-7

Reading Strategies

Use A Diagram

Quadrilaterals are plane figures that have four sides and four angles. All the figures inside the largest region are quadrilaterals.

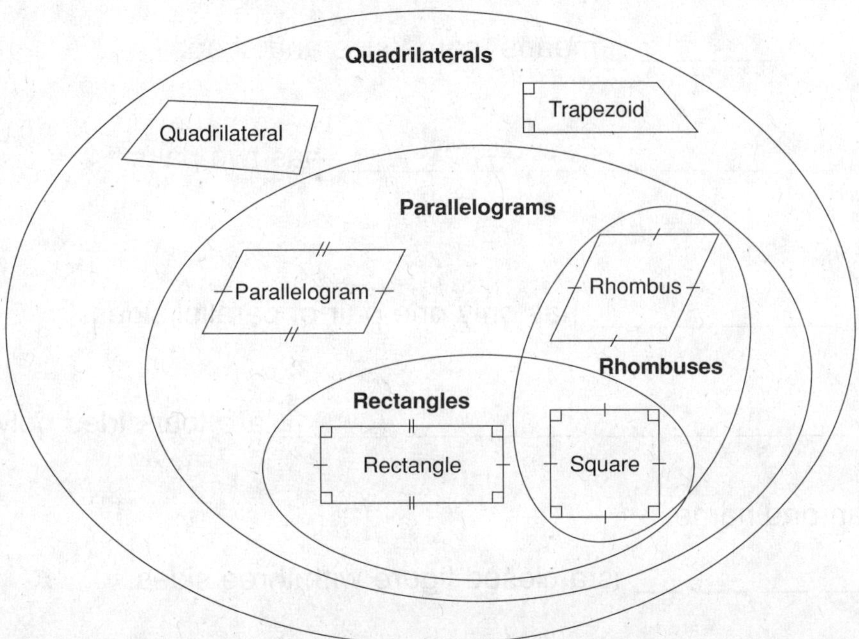

Parallelograms have two sets of opposite sides that are parallel and congruent. Opposite angles are also congruent. All the figures inside the next largest region are parallelograms.

Rectangles are quadrilaterals and parallelograms. They also have four right angles. **Rhombuses** are quadrilaterals and parallelograms with sides of equal length.

Use the diagram above to help you complete each question.

1. All parallelograms are quadrilaterals, true or false? ______________

2. Identify a quadrilateral that is not a parallelogram. ______________

3. Identify a quadrilateral that is also a parallelogram.. ______________

____

4. All parallelograms are rectangles, true or false? ______________

5. Identify a rectangle with four equal sides. ______________

6. A rhombus is a rectangle, true or false? ______________

Name ______________________ Date __________ Class __________

LESSON **8-7**

Puzzles, Twisters & Teasers

Charge!

Fill in the blanks to complete each statement. Use the letters above each number to solve the riddle below.

___ ___ ___ ___ ___ ___ ___ ___ ___ means "same size and shape."
1

A ___ ___ ___ ___ ___ ___ ___ ___ ___ ___ ___ ___ ___ has two pairs
2
of parallel sides.

A ___ ___ ___ ___ ___ ___ ___ ___ ___ has only one pair of parallel sides.
3

___ ___ ___ ___ ___ ___ ___ ___ ___ ___ ___ ___ ___ ___ are four-sided polygons
4
that can have more than one name.

A ___ ___ ___ ___ ___ ___ ___ ___ is a closed figure with three sides.
5

A ___ ___ ___ ___ ___ ___ ___ ___ ___ has four right angles.
6

A ___ ___ ___ ___ ___ ___ is a figure without angles or straight lines.
7

A ___ ___ ___ ___ ___ ___ has four congruent sides and four right angles.
8

A ___ ___ ___ ___ ___ ___ ___ has four congruent sides and no right angles.
9

Sides that are ___ ___ ___ ___ ___ ___ ___ ___ have a shared vertex.
10

How do you stop a rhinoceros from charging?

Take away his ___ ___ ___ ___ ___ ___   ___ ___ ___ ___
1 2 3 4 5 6   7 8 9 10

Name ______________________ Date ____________ Class ____________

LESSON 8-8

Practice A

Angles in Polygons

Find the unknown angle measure in each polygon. Choose the letter for the best answer.

1.

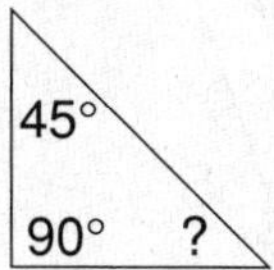

A 45° **C** 90°

B 55° **D** 135°

2.

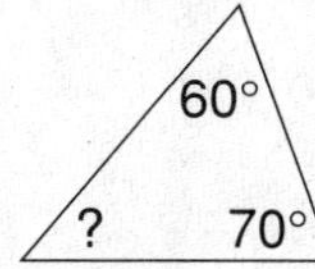

F 40° **H** 60°

G 50° **J** 70°

3.

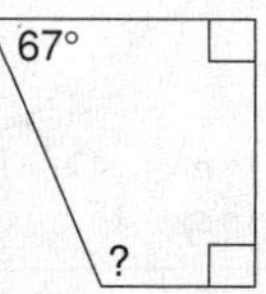

A 313° **C** 113°

B 170° **D** 27°

4.

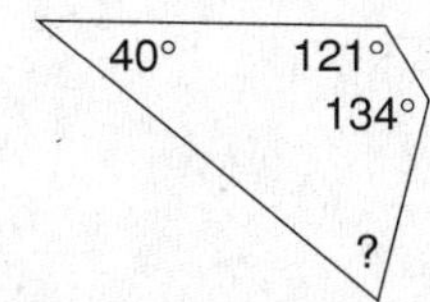

F 46° **H** 140°

G 65° **J** 245°

Find the unknown angle measure in each polygon.

5.

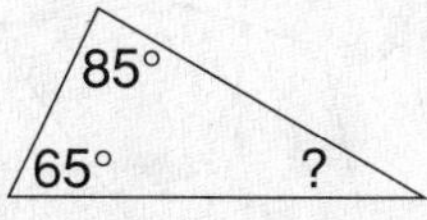

____

6.

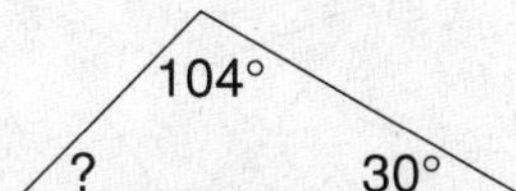

____

7.

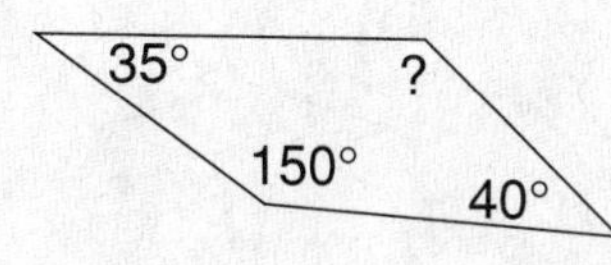

____

Divide each polygon into triangles to find the sum of its interior angles.

8.

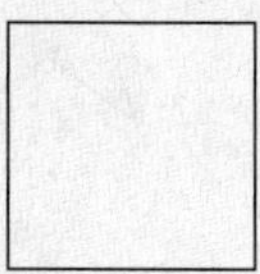

____

9.

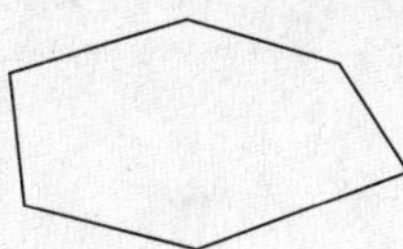

____

10.

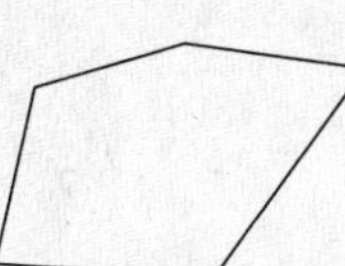

____

11.

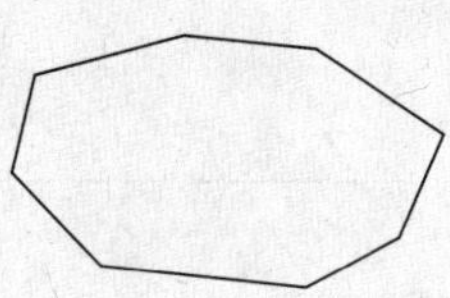

____

12.

____

13.

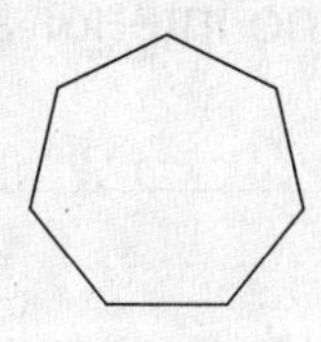

____

Name ______________________ Date __________ Class __________

LESSON 8-8

Practice B

Angles in Polygons

Find the unknown angle measure in each polygon.

1.

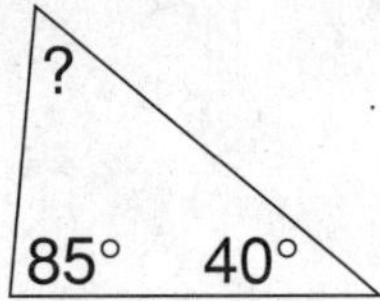

____

2.

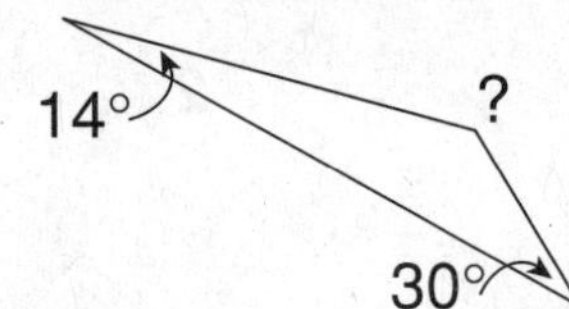

____

3.

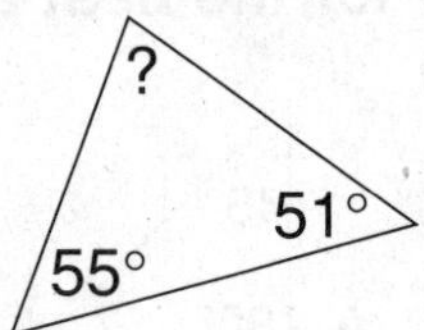

____

4.

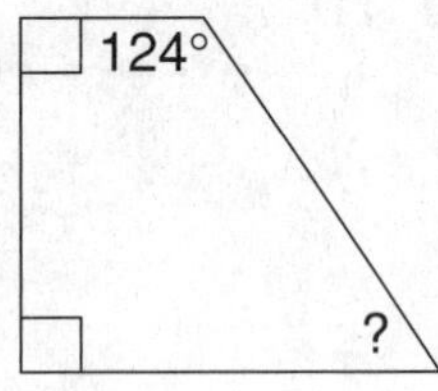

____

5.

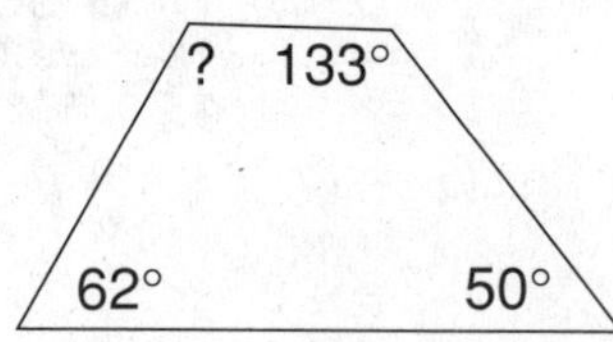

____

6.

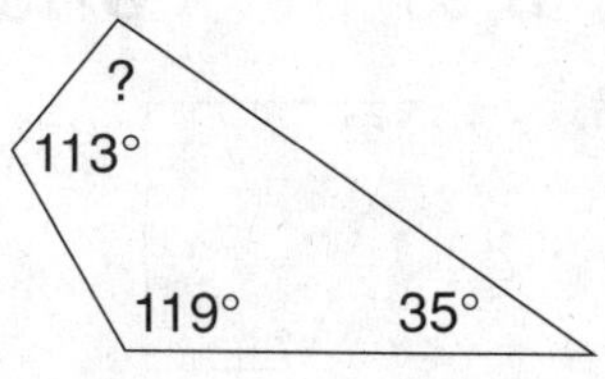

____

Divide each polygon into triangles to find the sum of its interior angles.

7.

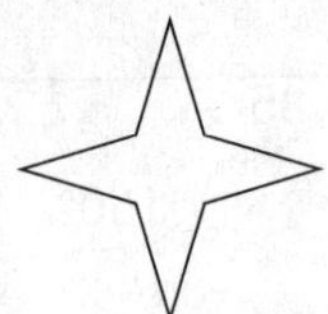

____

8.

____

9.

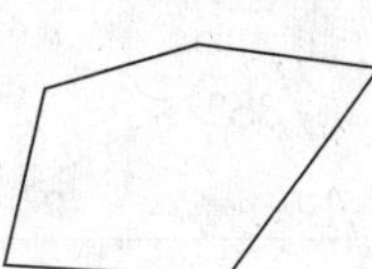

____

10.

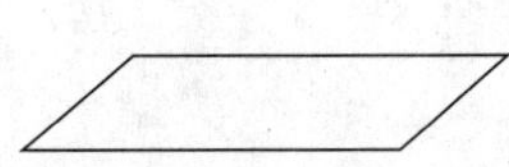

____

11.

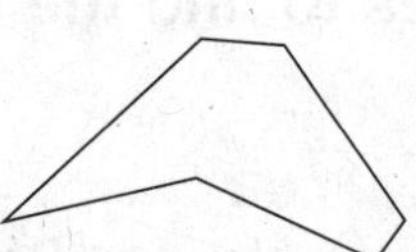

____

12.

____

13. A stop sign has the shape of a regular octagon. What is the sum of the interior angles of a stop sign?

____

Name ______________________ Date __________ Class __________

LESSON 8-8

Practice C

Angles in Polygons

Find the unknown angle measure in each polygon.

1.

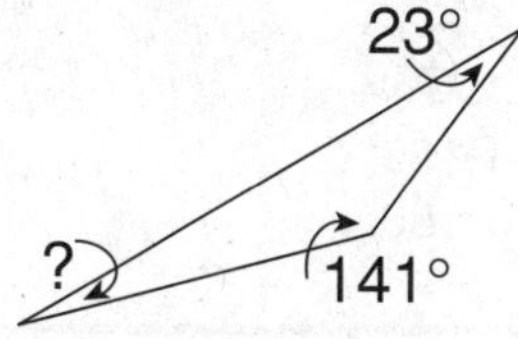

____

2.

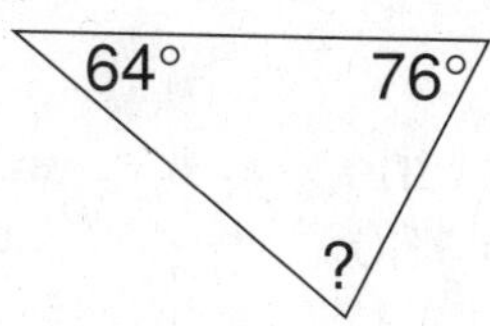

____

3.

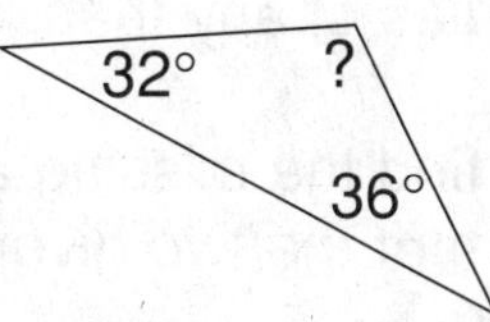

____

4.

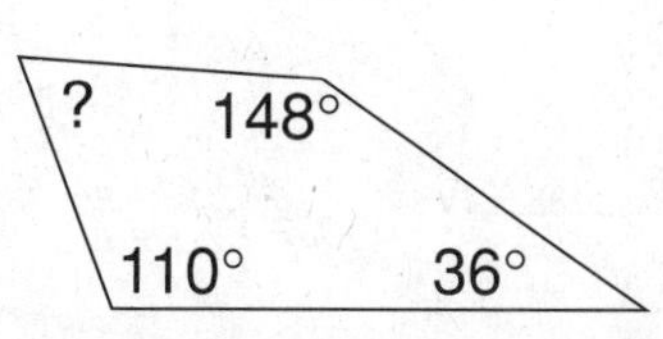

____

5.

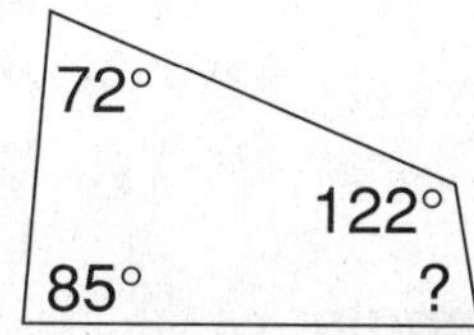

____

6.

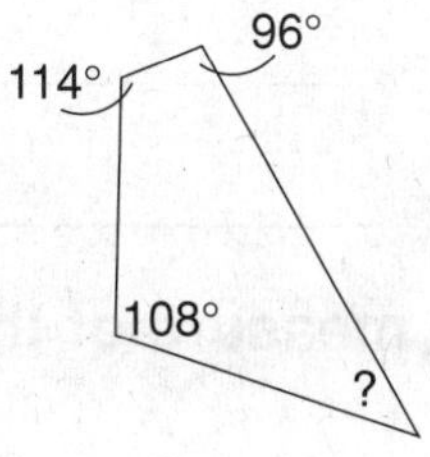

____

Divide each polygon into triangles to find the sum of its interior angles.

7.

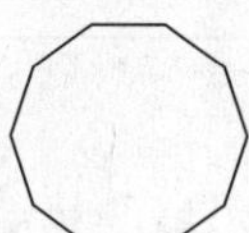

____

8.

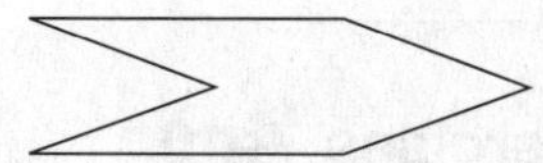

____

9.

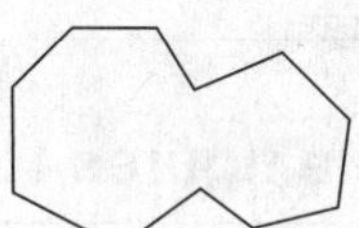

____

Find the measure of the third angle in each triangle, given two angle measures. Then classify the triangle.

10. 99°, 34°

____

11. 12°, 90°

____

12. 44°, 66°

____

13. Mitsue has 3 planks that are each 4 feet long. If she uses the planks to form a triangular flower bed, what will be the measure of each angle in the triangle?

____

Name ____________________ Date __________ Class __________

LESSON 8-8

Reteach

Angles in Polygons

The sum of the measures of the three angles of any triangle is 180°.

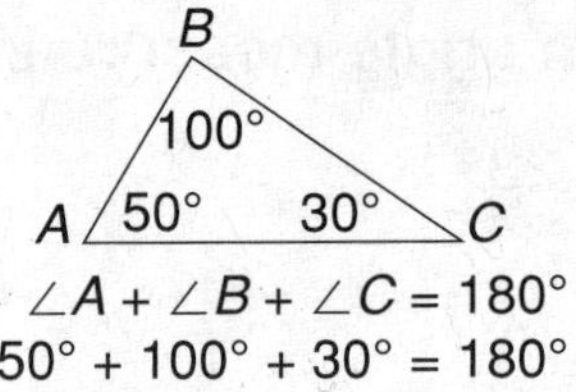

To find the missing angle, subtract the sum of the two given angles from 180°.

$\angle A + \angle B + \angle C = 180°$

$50° + 100° + 30° = 180°$

Find the measure of $\angle N$ in triangle LMN.

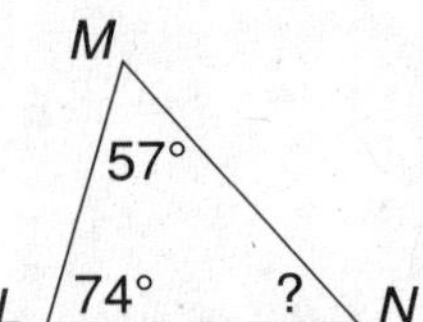

1. $\angle N =$ __________ $- (74° + 57°)$

2. $\angle N =$ __________ $- 131°$

3. $\angle N =$ __________

Find the measure of the unknown angle.

4. 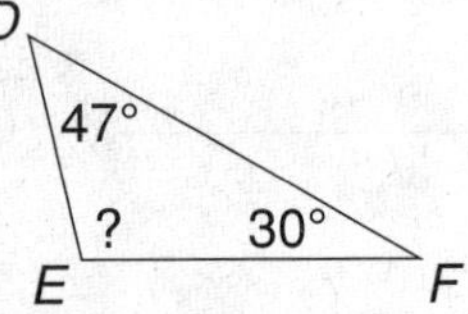

____

5. 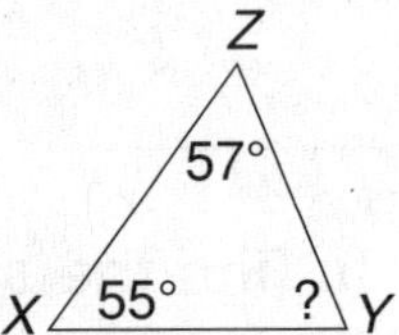

____

6. 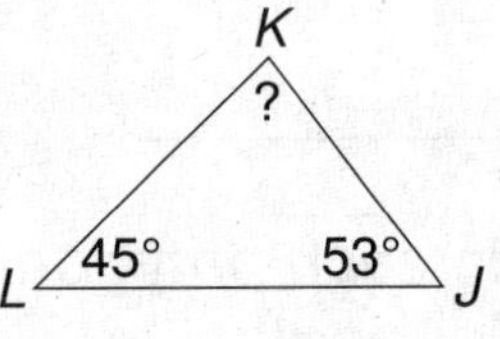

____

Use the figures to complete Exercises 7–10.

	Figure	Number of Sides	Number of Triangles
7.	Quadrilateral	4	
8.	Pentagon	5	
9.	Hexagon	6	

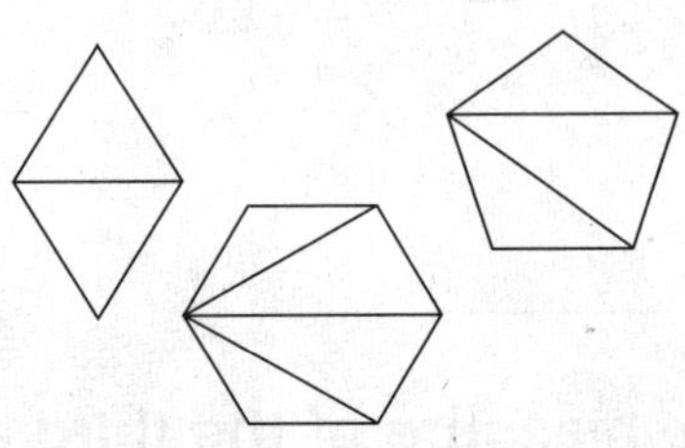

10. The number of triangles is always __________ less than the number of sides of the figure.

Find the sum of the interior angles.

11. 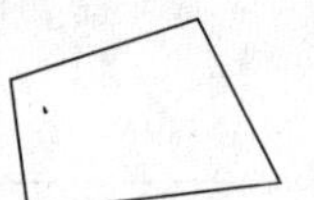

$180° \cdot$ _____ = _____

12. 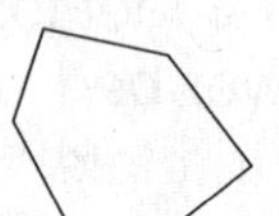

$180° \cdot$ _____ = _____

13. 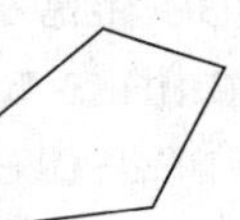

$180° \cdot$ _____ = _____

Name ______________________________ Date ____________ Class ____________

LESSON 8-8

Challenge

Shape Up

Draw a polygon based on each description. Label the measures of all angles. Name the polygon.

1. A figure has vertices *A, B, C, D,* and *E.* Sides *AB* and *BC* are 8 units in length. $\angle A = \angle C = 130°$. Side *CD* is parallel to *AE,* and each side is 10 units in length. $\angle D = \angle E = 90°$

____

2. A figure has vertices *W, X, Y,* and *Z.* Sides *WZ, WX,* and *XY* are each 12 units in length. Sides *WX* and *YZ* are parallel but are not the same length. $\angle W = \angle X = 120°$

____

3. A figure has vertices *M, N, O, P, Q,* and *R.* Sides *MN* and *QP* are parallel, and each is 15 units in length. $\angle M = \angle N = \angle P = \angle Q = 150°$

____

4. A figure has vertices *A, B, C,* and *D.* $AB = BC = CD = AD$. Side *AB* is parallel to *CD,* and side *AD* is parallel to *BC.* $\angle B$ and $\angle D = 125°$

____

5. A figure with vertices *E, F,* and *G* has perpendicular sides *EF* and *FG.* $\angle E = \angle G$

____

6. A figure has vertices *Q, R, S, T.* Side *QR* is parallel to *ST,* and side *QT* is parallel to *RS.* $\angle Q = 113°$

____

Name ______________________ Date __________ Class __________

LESSON 8-8

Problem Solving

Angles in Polygons

Write the correct answer.

These are some common street signs.

1. The interior angles of which sign have a sum of 360°?

____

2. What is the sum of the interior angles of a stop sign?

____

3. The interior angles of which sign have a sum of 540°?

____

4. If a yield sign is a regular polygon, what is the measure of each angle?

____

Choose the letter for the best answer.

5. A triangle with 3 unequal sides is a scalene triangle. Scalene triangles also have 3 unequal angles. Which angles can form a scalene triangle?

A 50°, 60°, 65°

B 50°, 60°, 70°

C 110°, 40°, 40°

D 20°, 20°, 40°

6. A design on a belt shows a row of irregular pentagons. Inside each figure, two of the angles each measure 60°. Another angle measures 270°. If the remaining angles are equal, what are their measures?

F 60° each **H** 75° each

G 120° each **J** 150° each

7. What is the missing angle measure of a triangle with angles 62° and 72°?

A 52° **C** 54°

B 34° **D** 46°

8. On a door is a sign that says "Welcome To Our Home." The sign is in the shape of a regular octagon. What is the measure of one of the interior angles of the sign?

F 135° **H** 80°

G 65° **J** 150°

Name ______________________ Date __________ Class __________

LESSON 8-8

Reading Strategies

Make Predictions

Angles inside a polygon are called **interior angles.**

The sum of the interior angles of any triangle always equals 180°.

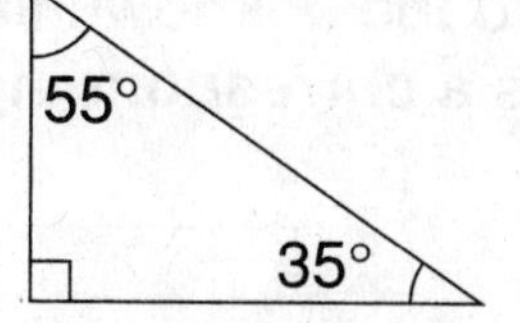

$35° + 55° + 90° = 180°$

To find the sum of interior angles for other polygons:

- Divide the polygon into triangles.
- Draw all possible diagonals from one vertex.
- Multiply the number of triangles formed by 180°.

This quadrilateral can be divided into two triangles.

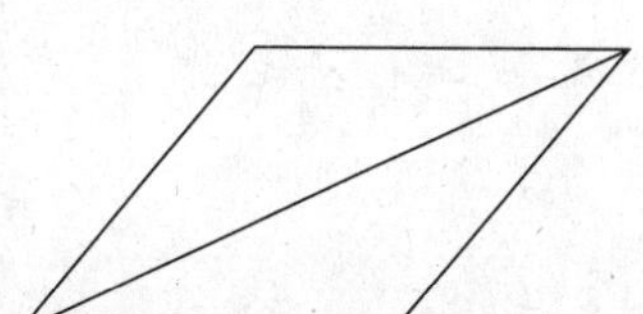

2 triangles formed

This pentagon can be divided into three triangles.

3 triangles formed

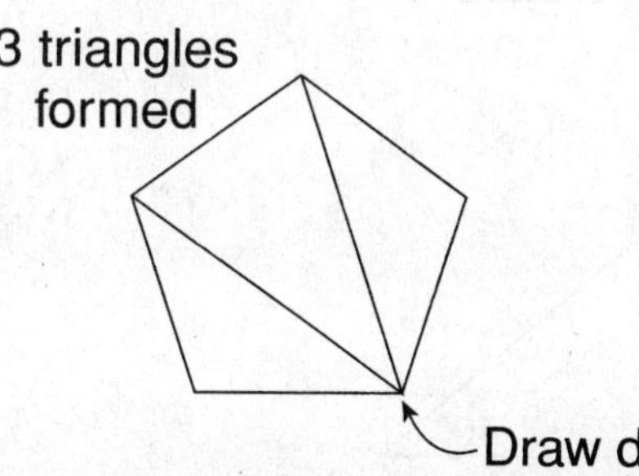

Draw diagonals from same vertex

You can make a list of polygons and the number of triangles that can be made inside them.

Polygon	Number of Sides	Number of Triangles
Triangle	3	1
Quadrilateral	4	2
Pentagon	5	3
Hexagon	6	?
Septagon	7	?
Octagon	8	?

Use the pattern in the list to predict the number of triangles for the rest of the polygons.

1. How many triangles do you think can be formed inside a hexagon? __________

2. What helped you make your prediction?

____

3. How many triangles do you predict can be formed inside a septagon? __________

4. How many triangles do you predict can be formed inside an octagon? __________

Name ______________________ Date __________ Class __________

LESSON **8-8**

Puzzles, Twisters & Teasers

What's Your Angle?

Find the unknown angle measure in each figure. Each answer has a corresponding letter. Use the letters to solve the riddle.

1. $t =$ ____________

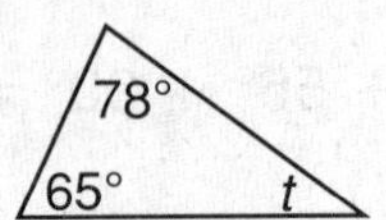

2. $n =$ ____________

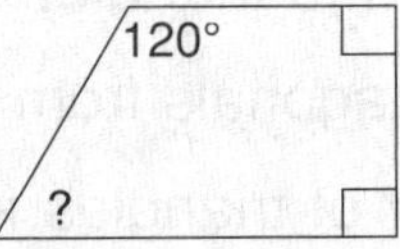

3. $i =$ ____________

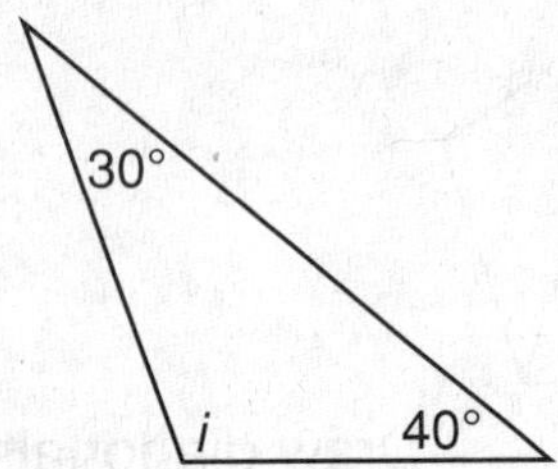

4. $a =$ ____________

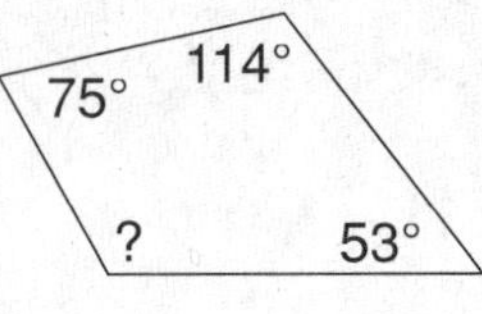

5. $c =$ ____________

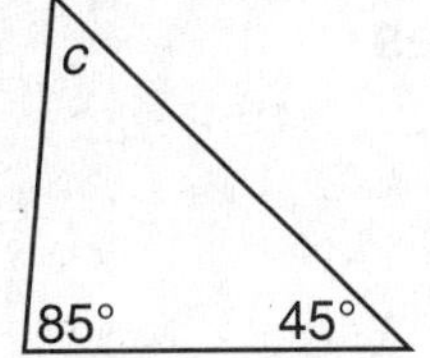

6. $u =$ ____________

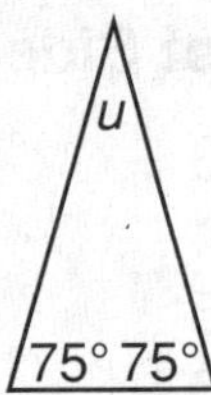

Why would you take a pencil to bed?

To draw the ___ ___ **R** ___ ___ ___ ___ **S**

50 30 37 118 110 60

Name ____________________ Date __________ Class __________

LESSON 8-9

Practice A

Congruent Figures

Are there any congruent figures in each picture? If there are, describe them.

1.

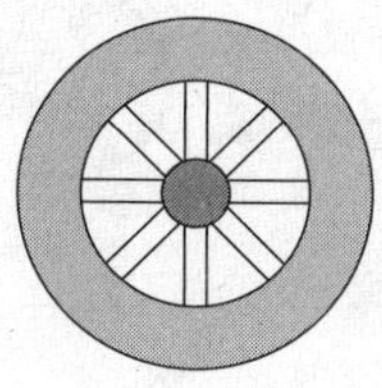

____

____

2.

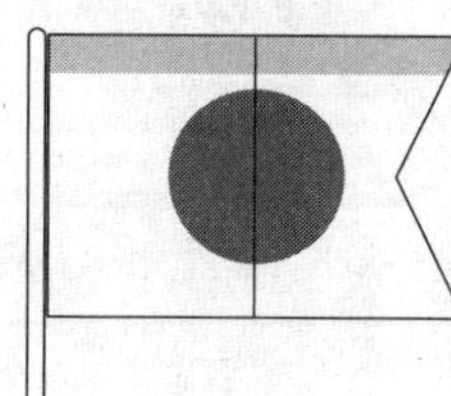

____

____

3.

4 ft 5 ft 7 ft 4 ft 5 ft 7 ft

____

4.

12 cm 4 cm 9 cm 12 cm 4 cm 9 cm

____

5.

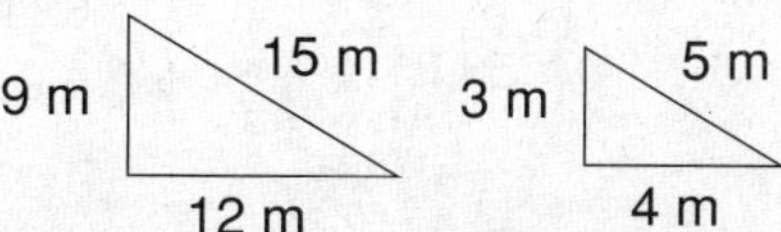

____

6.

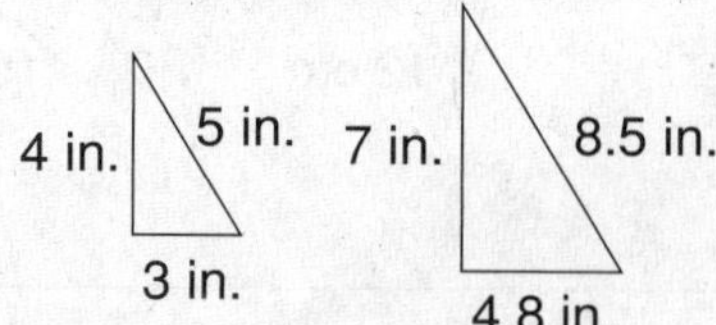

____

Determine the missing measure in each set of congruent polygons.

7.

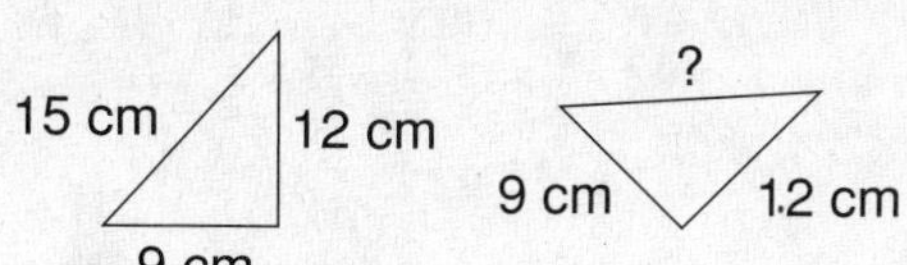

____

8.

7 in. 82° 4 in. ? 60° 8 in. 4 in. 82° 7 in. 60° 38° 8 in.

____

9.

7.5 m, 4 m, 105°, 75°, 75°, 105°, 4 m, 7.5 m 7.5 m, 4 m, ?, 75°, 75°, 105°, 4 m, 7.5 m

____

10.

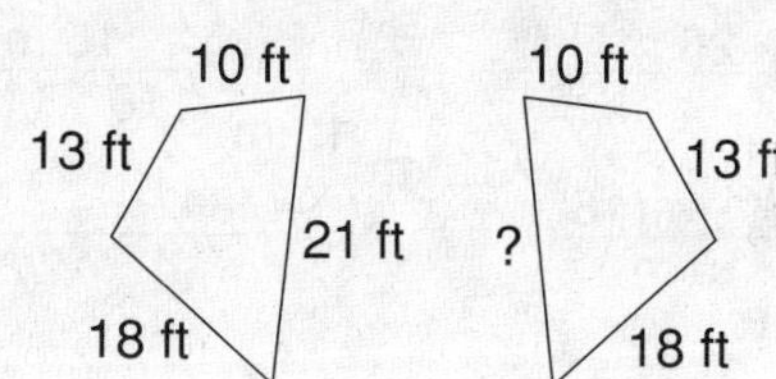

____

Name ______________________ Date ____________ Class ____________

LESSON 8-9

Practice B

Congruent Figures

Identify any congruent figures.

1.

2.

Determine whether the triangles are congruent.

3.

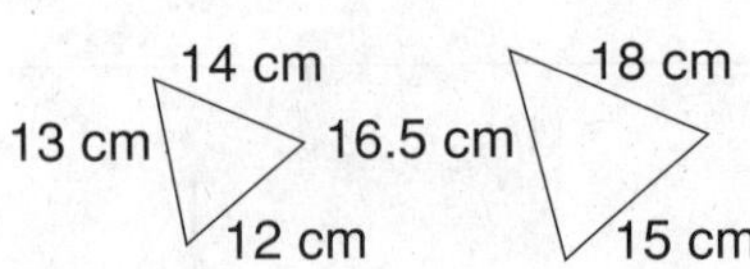

22 m, 10 m, 14 m — 14 m, 10 m, 22 m

4.

3 in., 3 in., 5 in. — 7 in., 4 in., 4 in.

5.

14 cm, 13 cm, 12 cm, 16.5 cm — 18 cm, 15 cm

6.

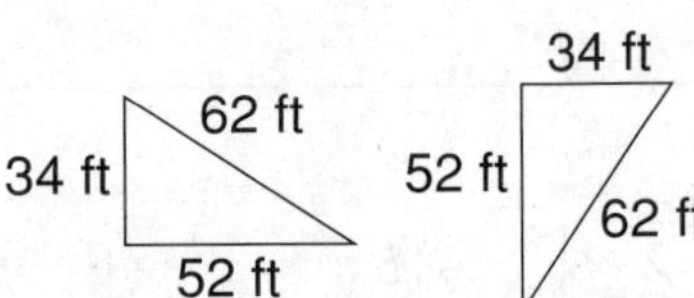

Determine the missing measure or measures in each set of congruent polygons.

7.

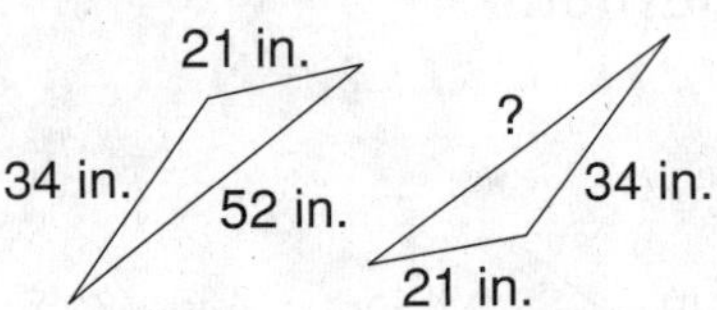

8.

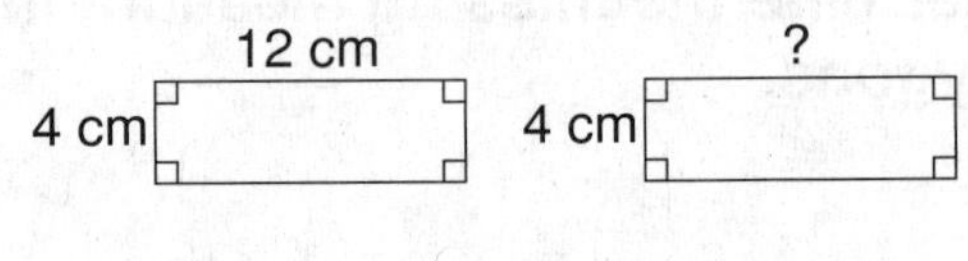

9.

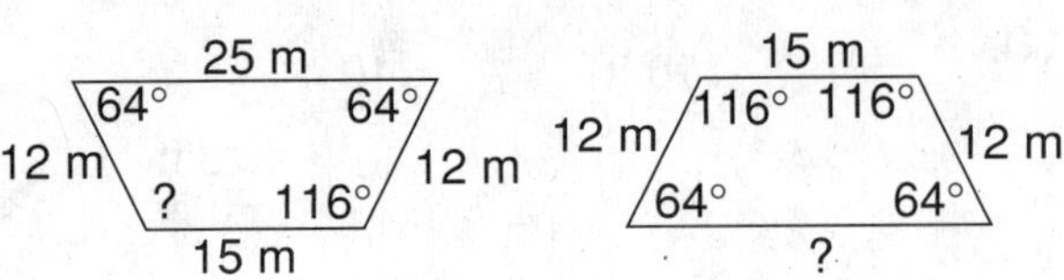

10.

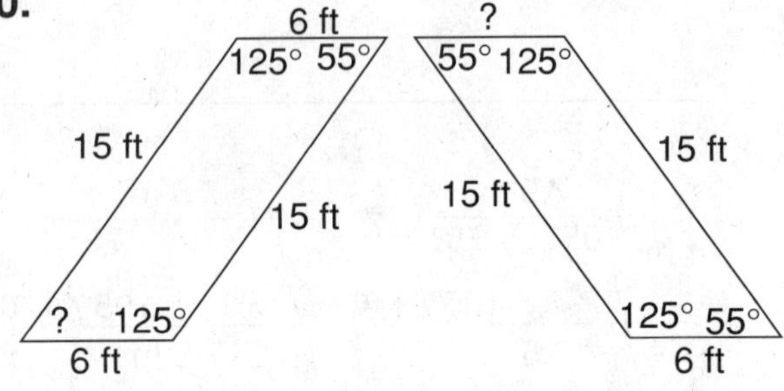

Name ______________________ Date __________ Class __________

LESSON 8-9

Practice C

Congruent Figures

Identify any congruent figures.

1.

____

____

2.

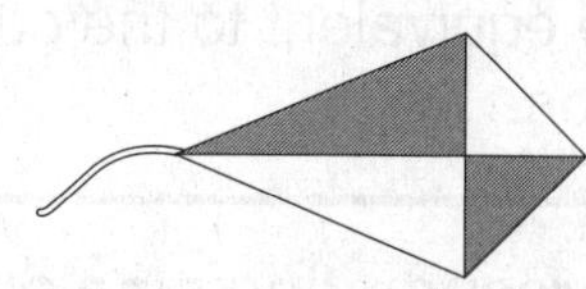

____

____

____

Determine whether the triangles are congruent.

3.

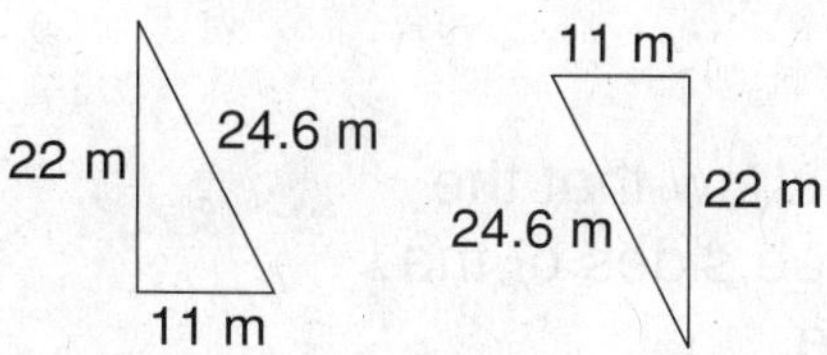

____

4.

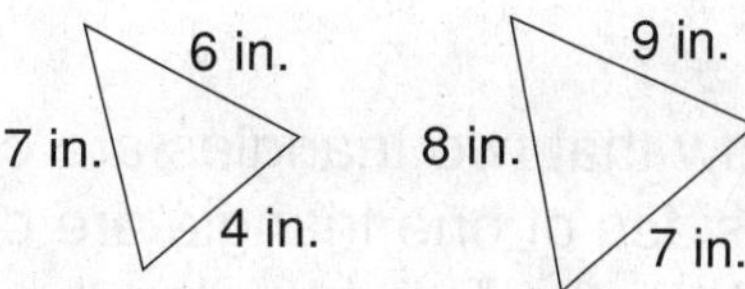

____

Determine the missing measures in each set of congruent polygons.

5.

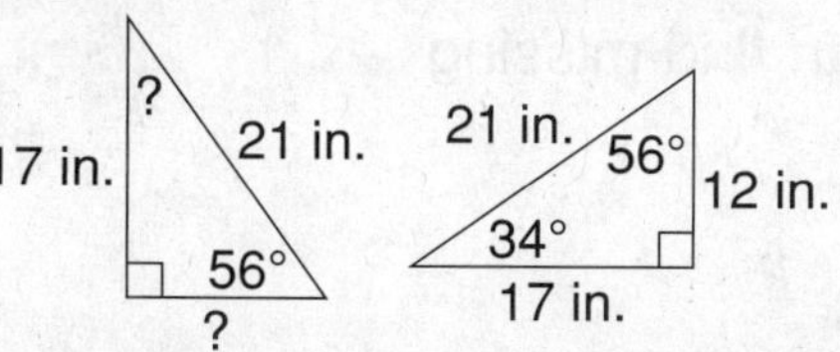

____

6.

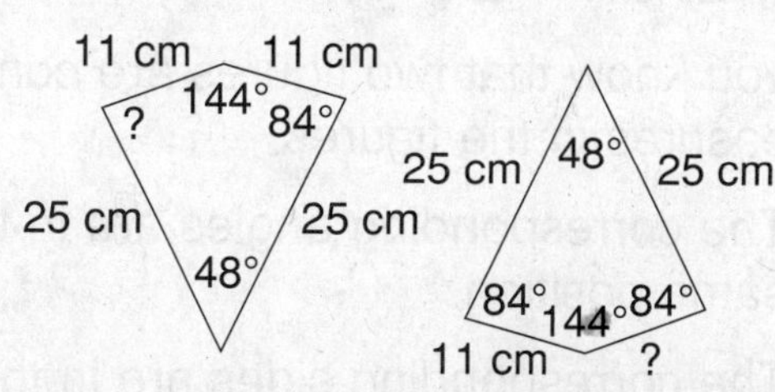

____

7.

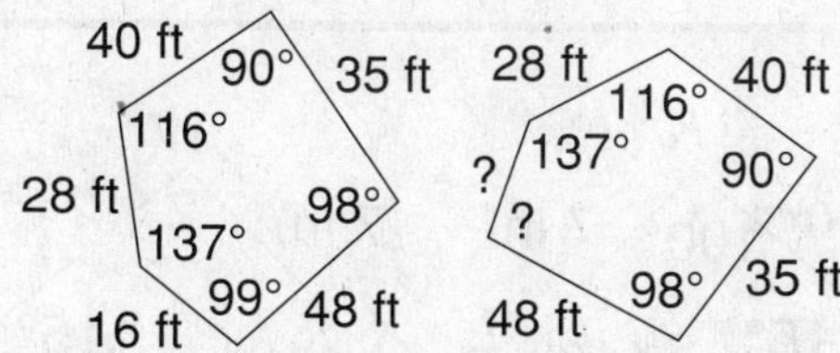

____

8.

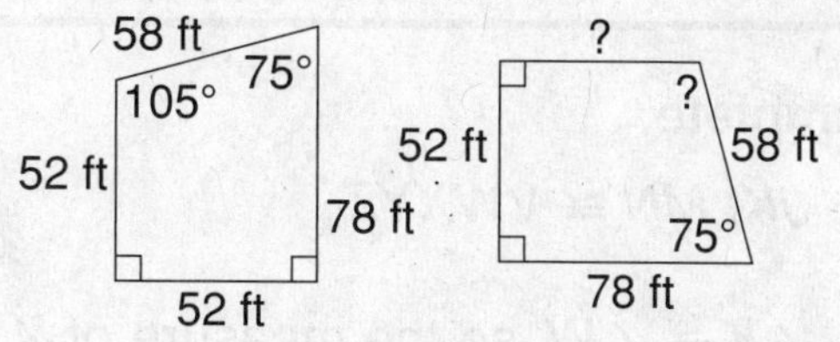

____

9. Describe how to determine if two hexagons are congruent.

____

____

Name ______________________________ Date ____________ Class ____________

LESSON **8-9**

Reteach

Congruent Figures

When two polygons are congruent, the angles and the sides of one polygon are equivalent to the corresponding angles and sides of the other polygon.

Find the corresponding side or angle. $ABCD \cong EFGH$

1. $\overline{AB} \cong$ ____________

2. $\angle DAB \cong$ ____________

3. $\overline{GH} \cong$ ____________

4. $\angle GFE \cong$ ____________

A 9 cm B
6 cm 8 cm
D 14 cm C

F 9 cm E
8 cm 6 cm
G 14 cm H

To show that two triangles are congruent, you can show that the three sides of one triangle are congruent to the three sides of the other triangle. This is called the *Side-Side-Side* rule.

$\overline{PQ} \cong \overline{ST}$ since both are 8 inches.

$\overline{QR} \cong \overline{TU}$ since both are 6 inches.

$\overline{PR} \cong \overline{SU}$ since both are 10 inches.

So, $\triangle PQR \cong \triangle STU$.

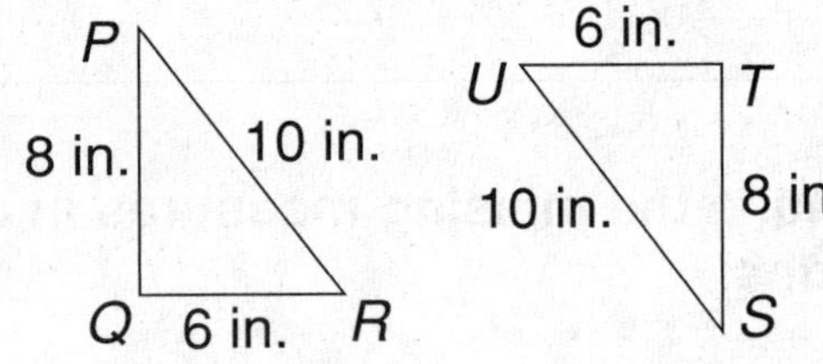

If you know that two figures are congruent, you can find missing measures in the figures.

- The corresponding angles are in the same position.
- The corresponding sides are in the same position.

Complete.

5. *JKLMN* $\cong$ *VWXYZ*

   $\angle K \cong \angle W$, so the measure of $\angle W$ is ____________.

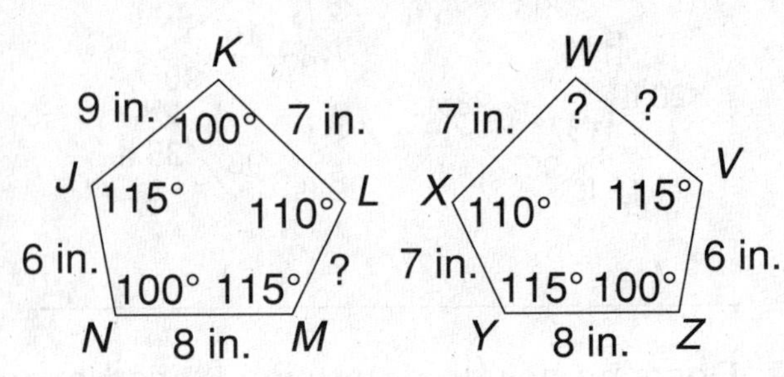

6. *JKLMN* $\cong$ *VWXYZ*

   $\overline{LM} \cong \overline{XY}$, so the length of $\overline{LM}$ is ____________

7. What is the length of $\overline{VW}$? ____________

Name ______________________________ Date ____________ Class ____________

LESSON **8-9**

Challenge

Congruent Presidents?

Find the figures in each row that appear to be congruent. Write their letters on the line next to the figures. Unscramble the letters to name the president.

1.

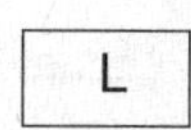

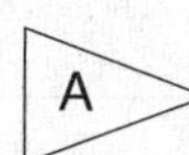

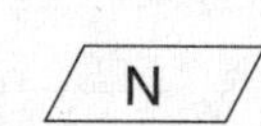

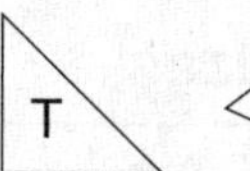

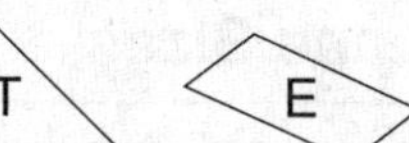

____

2.

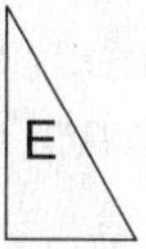

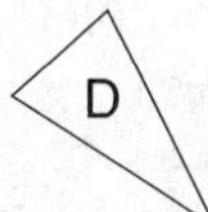

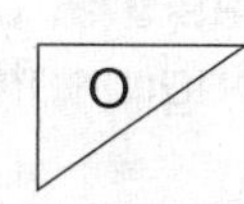

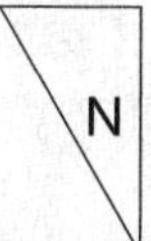

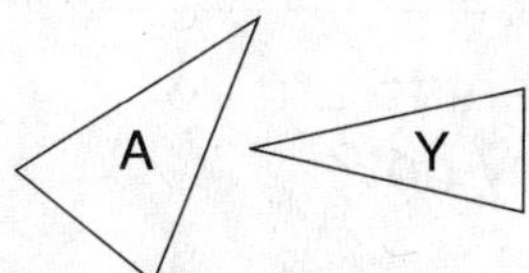

____

3. This president had a pony named Macaroni. ______________

4.

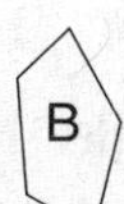

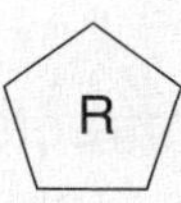

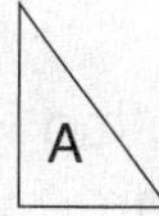

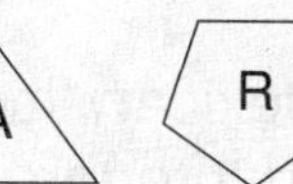

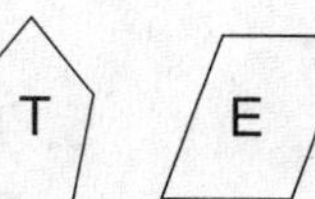

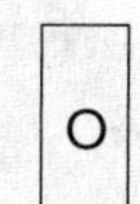

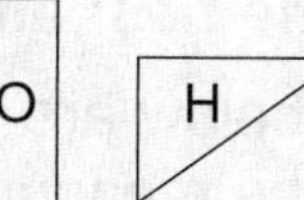

____

5.

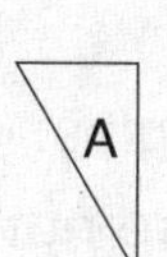

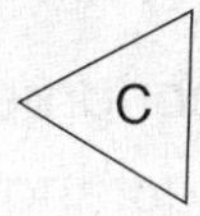

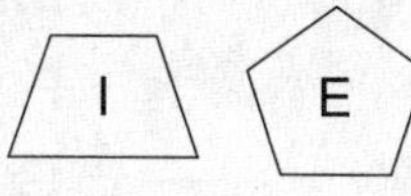

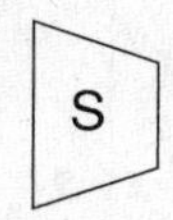

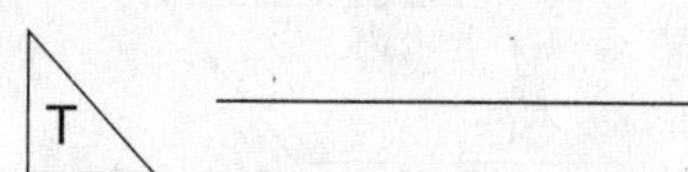

____

6. This president had a goat named Old Whiskers who used to pull the president's grandchildren in a cart. ______________

7.

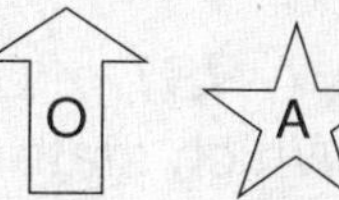

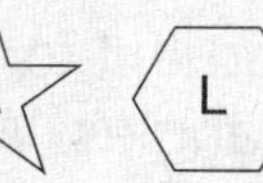

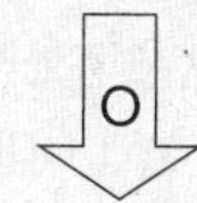

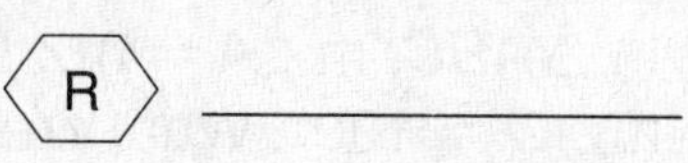

____

8.

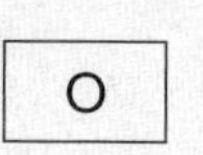

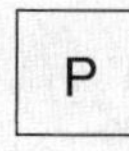

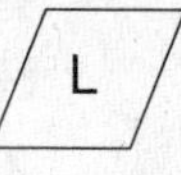

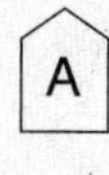

____

9. This president walked his raccoon named Rebecca on a leash at the White House. ______________

Name ______________________ Date __________ Class __________

LESSON 8-9

Problem Solving

Congruent Figures

Write the correct answer.

The table shows the dimensions of regulation NBA and NCAA basketball courts, which are rectangular in shape.

Basketball Court and Lane Sizes

	NBA	NCAA
Court	94 ft by 50 ft	94 ft by 50 ft
Lane	16 ft by 19 ft	12 ft by 19 ft

1. Is an NBA court congruent to an NCAA court? Why or why not?

____

____

2. If the lane on an NBA court is the same shape as the lane on an NCAA court, are they congruent? Explain.

____

____

3. The lane on a WNBA court is 12 feet by 19 feet. If the lane is the same shape as that of the NBA lane, are they congruent? Why or why not?

____

____

Choose the letter for the best answer.

4. The parallelograms below are congruent. What is the measure of $\angle U$?

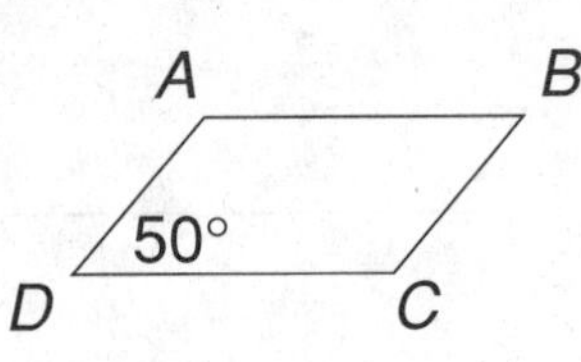

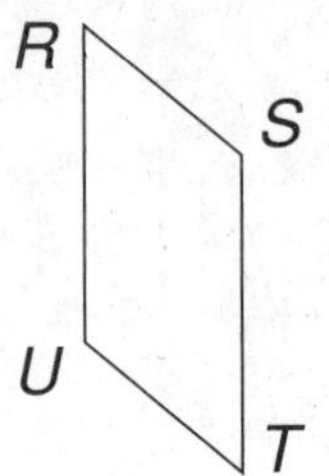

A 50° **C** 100°

B 130° **D** 260°

5. The Mexican flag is a rectangle divided into three vertical stripes of identical measures. Which of the following statements is true about the Mexican flag?

F It has 3 congruent rectangles.

G It has 3 rectangles that are not congruent.

H It has 4 congruent rectangles.

J It has 4 rectangles and none are congruent.

6. In $\triangle ABC$, $m\angle A = m\angle B$, and $m\angle C = 100°$. What are the measures of the angles in $\triangle DEF$ if it is congruent to $\triangle ABC$?

A 60°, 60°, 100°

B 20°, 60°, 100°

C 30°, 60°, 100°

D 40°, 40°, 100°

7. $\triangle JKL$ is congruent to $\triangle RST$. $\angle L$ and $\angle T$ are right angles. Which statement about the two triangles is *not* true?

F $m\angle K = m\angle S$

G $\overline{JK} \cong \overline{RS}$

H $m\angle J = m\angle L$

J $m\angle R + m\angle S = 90°$

Name ______________________________ Date ____________ Class ____________

LESSON **8-9**

Reading Strategies

Graphic Organizer

This chart helps you understand congruence.

Congruence

Definition

Two figures that have exactly the same size and shape are congruent.

Facts

- Corresponding sides are congruent.
- Corresponding angles are congruent.

Examples

T 37°
4 cm
5 cm
D 3 cm N

J 37°
5 cm
4 cm
M 3 cm B

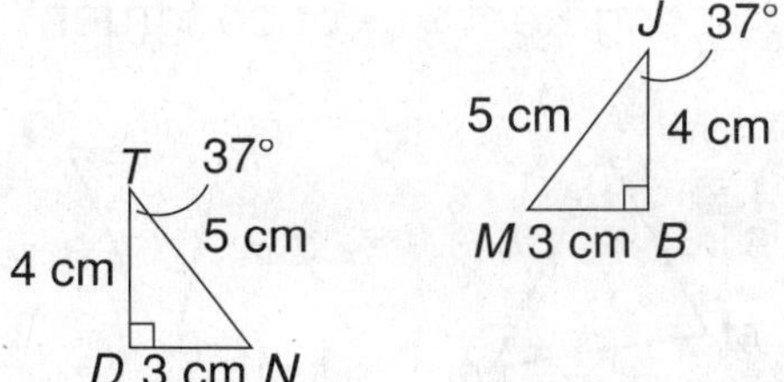

Non-examples

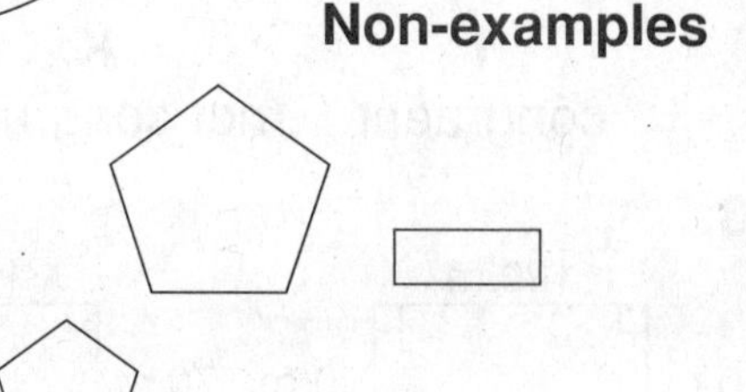

Use the chart to answer each question.

1. How can you tell if two figures are congruent?

____

2. Are the rectangles congruent? Why or why not?

____

3. Compare the two triangles. Which side corresponds to *TD*?

____

4. How long are $\overline{TD}$ and $\overline{JB}$?

____

5. Which angle corresponds to angle *D*?

____

6. Are the two triangles congruent? Explain why or why not.

____

____

Name ______________________ Date ____________ Class ____________

LESSON **8-9**

Puzzles, Twisters & Teasers

Hit the Nail on the Head!

Decide whether or not each pair of figures is congruent. Circle the letter above your answer. Unscramble the letters to solve the riddle.

1.

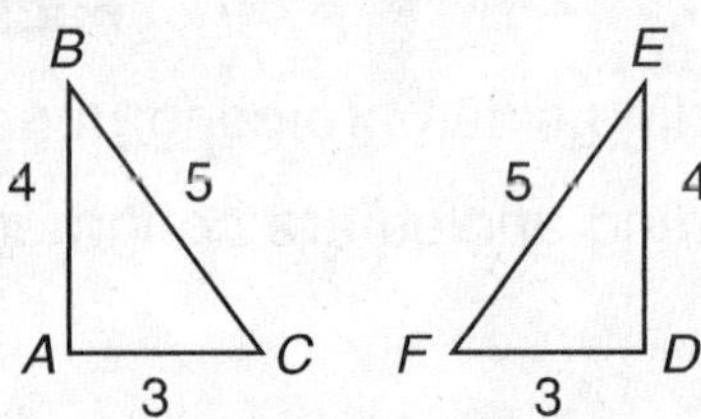

T	**K**
congruent	not congruent

2.

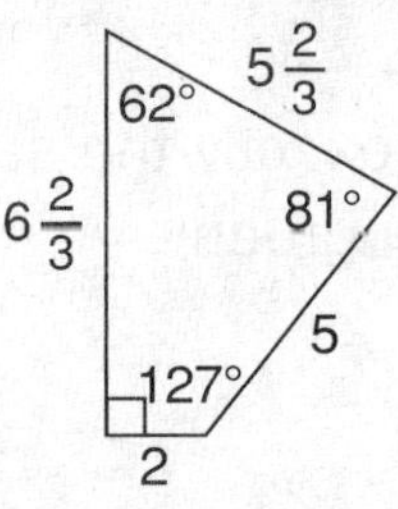

P	**H**
congruent	not congruent

3.

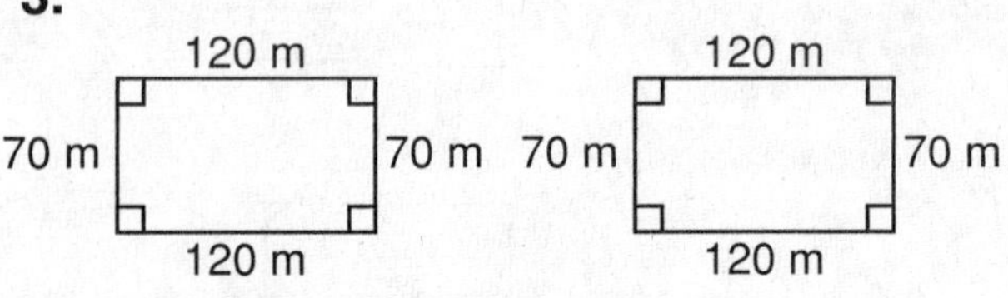

M	**R**
congruent	not congruent

4.

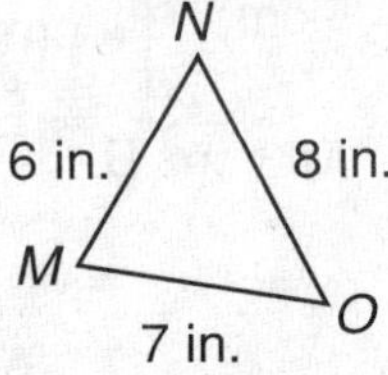

L	**A**
congruent	not congruent

5.

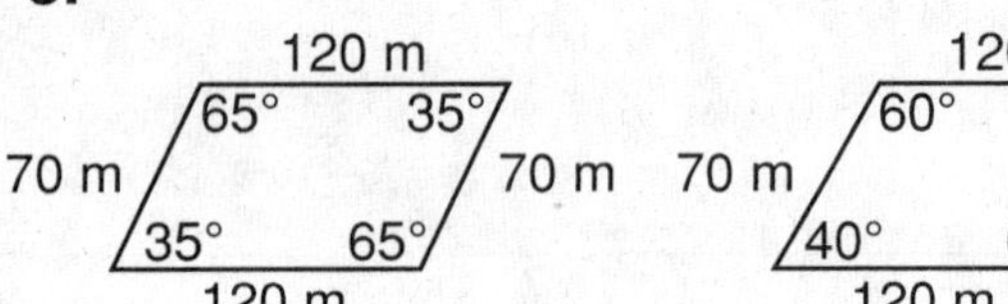

U	**B**
congruent	not congruent

6.

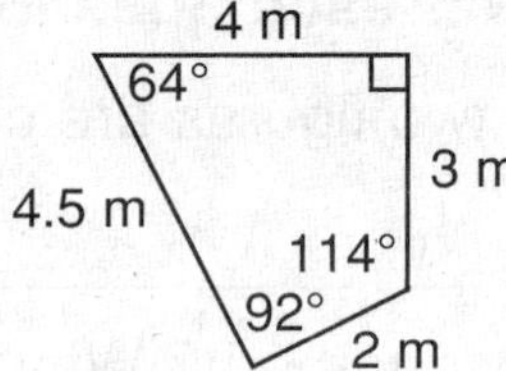

I	**W**
congruent	not congruent

Which nail does a carpenter hate to hit?

His ___ ___ **U** ___ ___ **N** ___ ___ **L**

Name ________________________ Date ____________ Class ____________

LESSON **8-10**

Practice A

Translations, Reflections, and Rotations

Identify the transformation. Choose the letter of the best answer.

1.

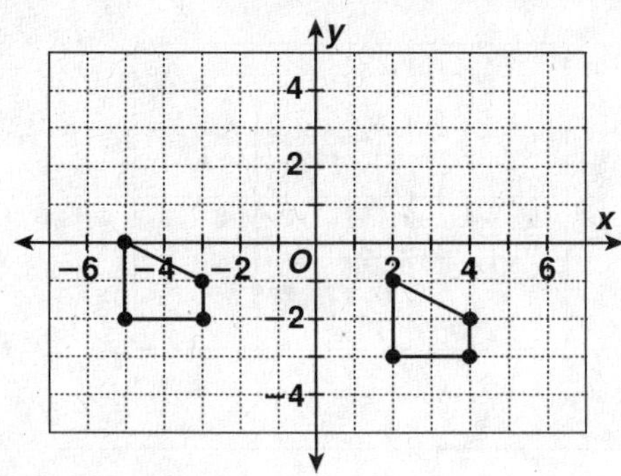

A translation
B reflection
C rotation

2.

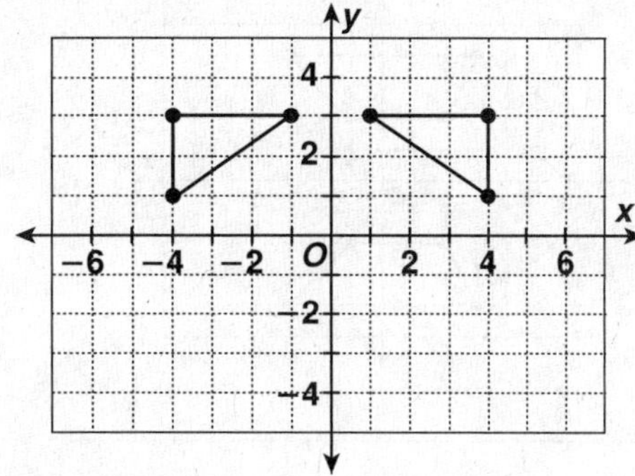

A translation
B reflection
C rotation

Graph each translation.

3. 2 units to the left and 4 units down

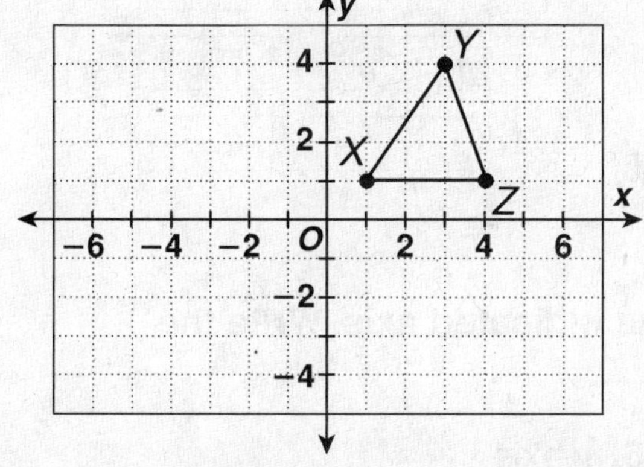

4. 3 units to the right and 2 units up

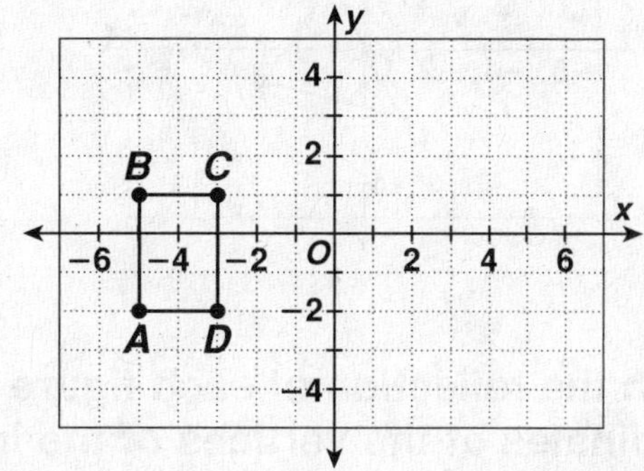

Follow the directions to graph each transformation.

5. Reflect $\triangle EFG$ across the x-axis.

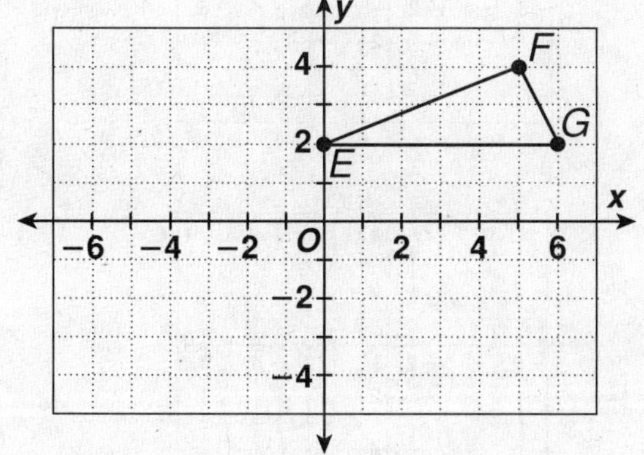

6. Rotate $\triangle XYZ$ 180° about the vertex X.

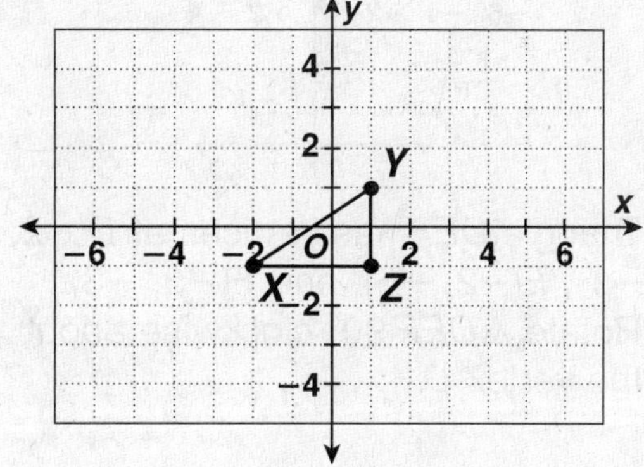

Name ______________________ Date __________ Class __________

LESSON 8-10

Practice B

Translations, Reflections, and Rotations

Identify each type of transformation.

1.

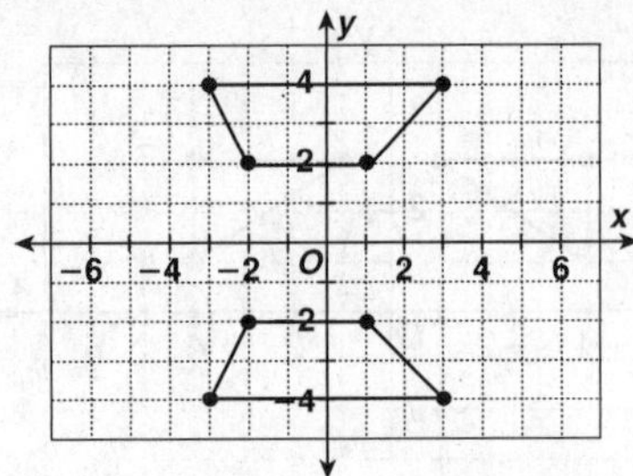

____

2.

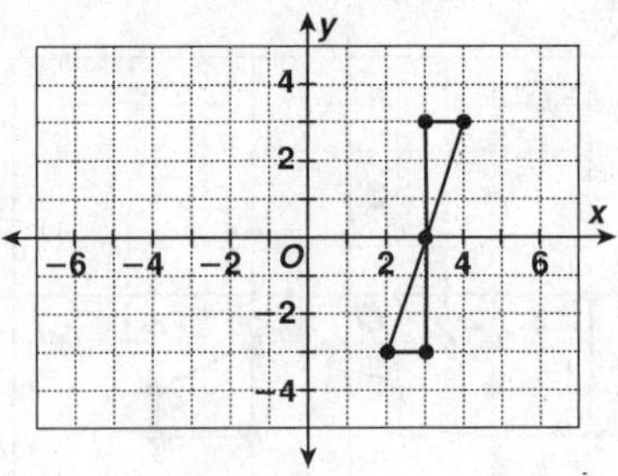

____

Graph each translation.

3. 5 units to the left and 2 units up

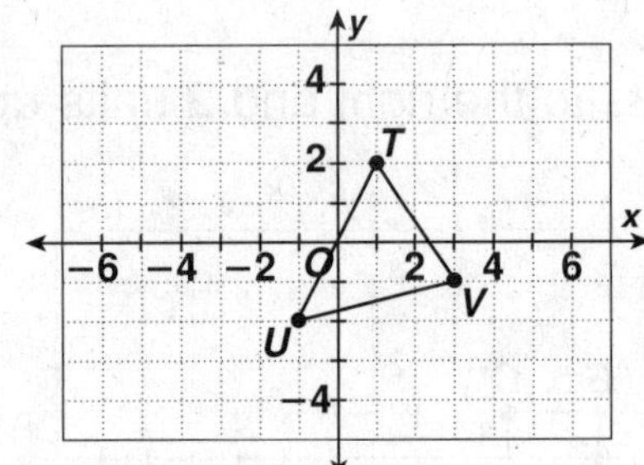

4. 4 units to the right and 3 units up

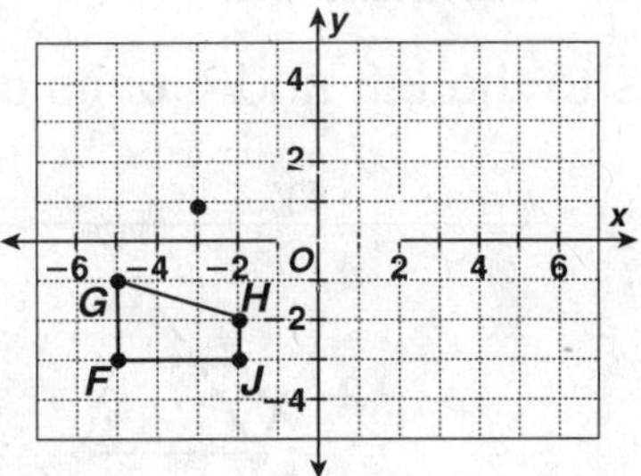

Graph the reflection of each figure across the indicated axis. Write the coordinates of the vertices of the image.

5. *x*-axis

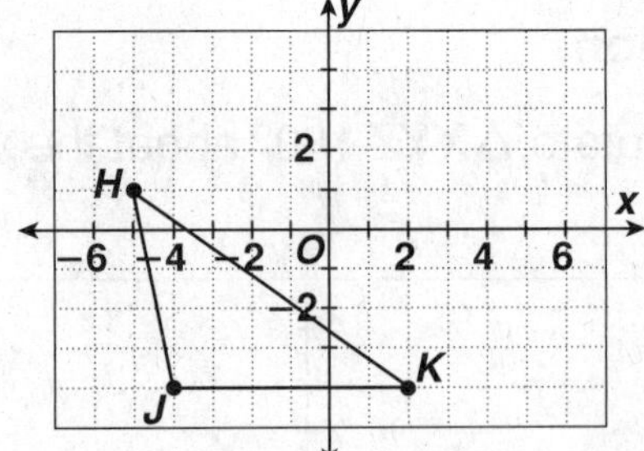

6. *y*-axis

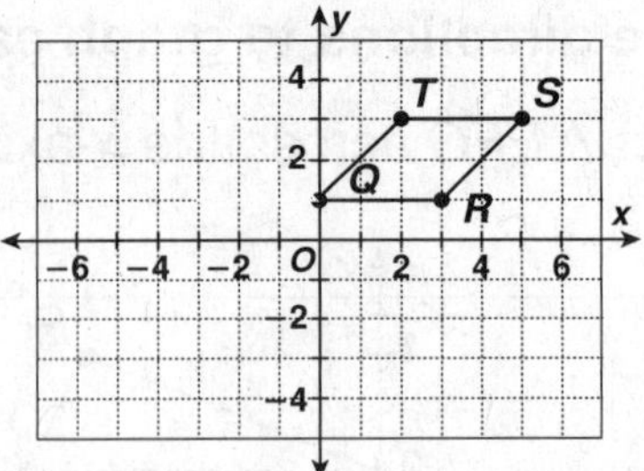

7. Triangle *DEF* has vertices at *D*(−2, −1), *E*(−2, −3), and *F*(−5, −3). Rotate $\triangle DEF$ 90° clockwise about the vertex *D*.

Name ______________________ Date __________ Class __________

LESSON 8-10

Practice C

Translations, Reflections, and Rotations

Identify each type of transformation.

1.

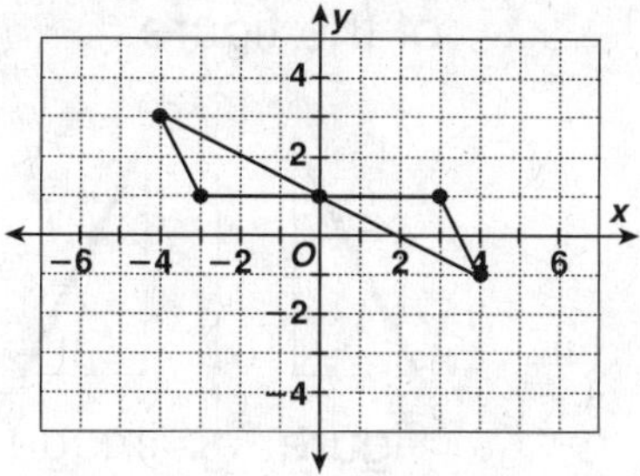

____

2.

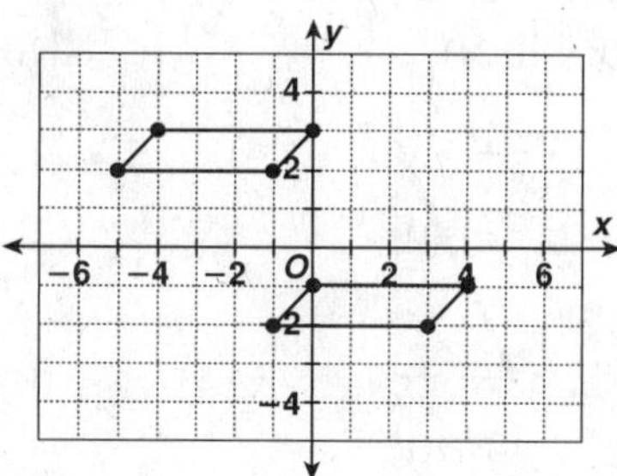

____

Graph each translation.

3. 3 units to the right and 3 units down

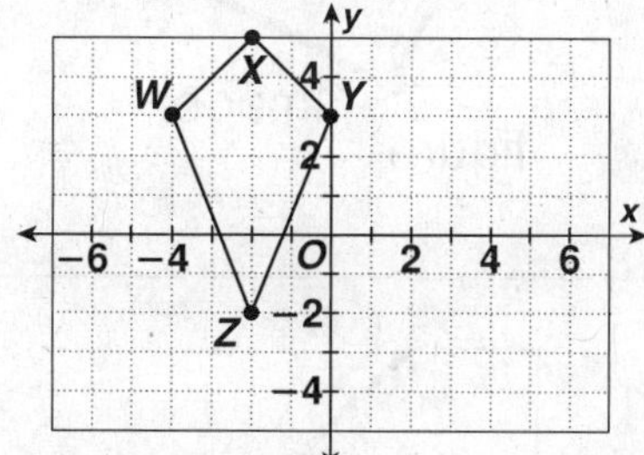

4. 4 units to the left and 5 units up

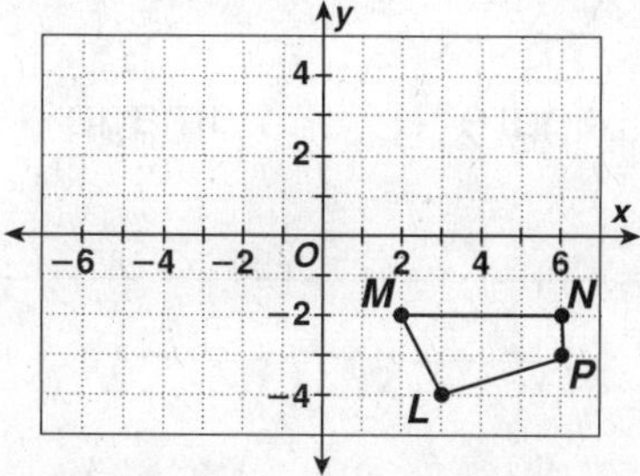

Graph the reflection of each figure across the indicated axis. Write the coordinates of the vertices of the image.

5. *x*-axis

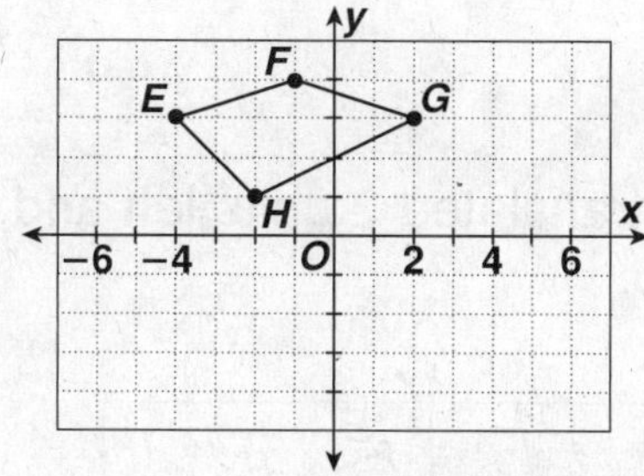

6. *y*-axis

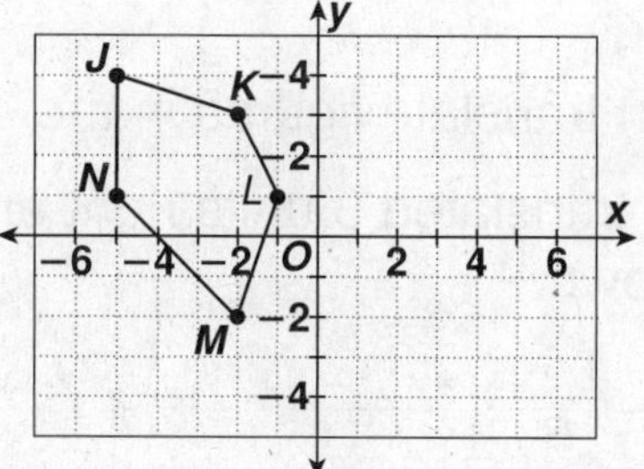

7. Triangle *KLM* has vertices at *K*(1, −1), *L*(4, −1), and *M*(5, 1). Rotate $\triangle KLM$ 270° counterclockwise about the vertex *K*.

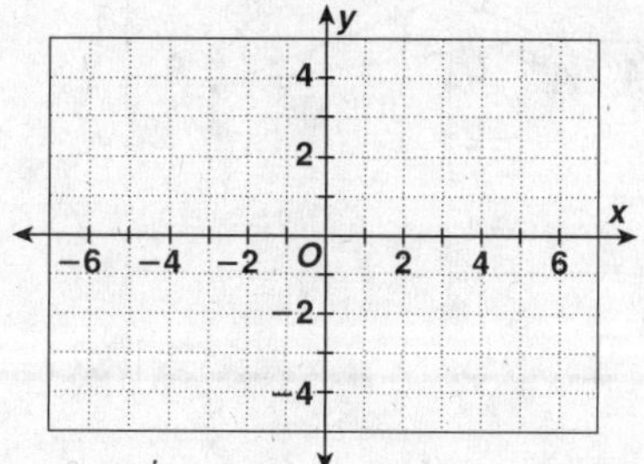

Name ______________________ Date __________ Class __________

LESSON 8-10

Reteach

Translations, Reflections, and Rotations

A **translation** is a *slide* to a new position.

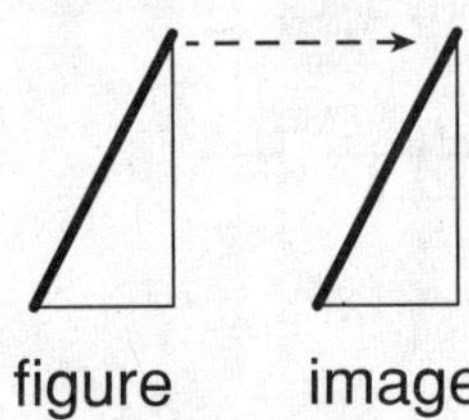

A **rotation** is a *turn* of the figure.

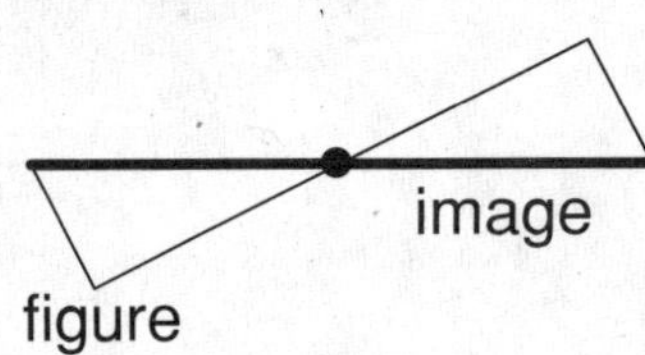

A **reflection** is a *flip* of the figure.

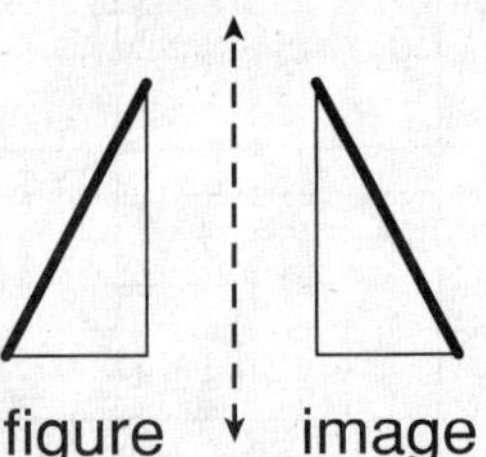

Identify the type of transformation.

1.

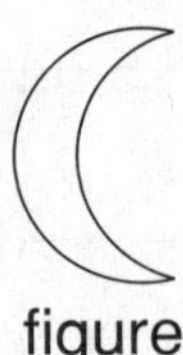

____

2.

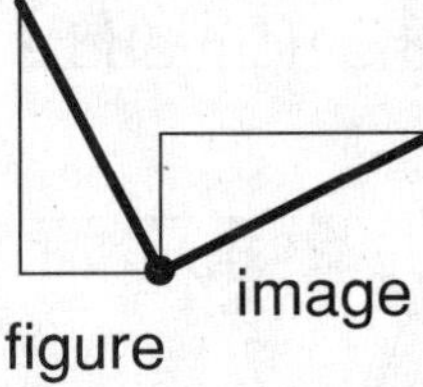

____

3.

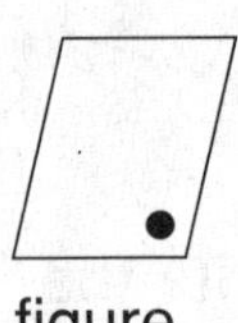

____

4.

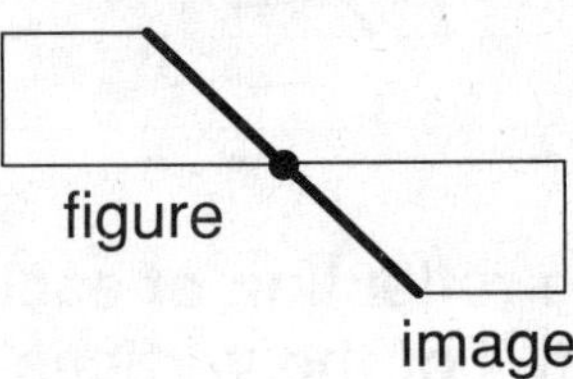

____

- You can translate figures in the coordinate plane.

$\triangle ABC$ is translated 3 units right and 3 units down.

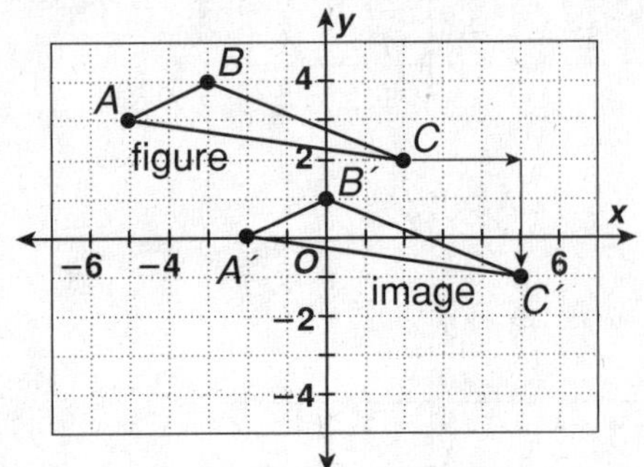

$\triangle DEF$ is translated 2 units left and 3 units up.

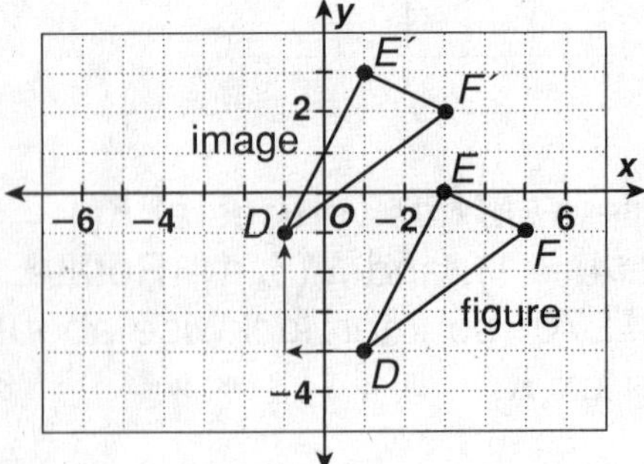

Name ______________________ Date __________ Class __________

LESSON 8-10

Reteach

Translations, Reflections, and Rotations (continued)

5. When you translate a figure to the right, does the x-coordinate of the image increase or decrease? __________

6. When you translate a figure downward, does the y-coordinate of the image increase or decrease? __________

- You can rotate a figure about a point in the coordinate plane.

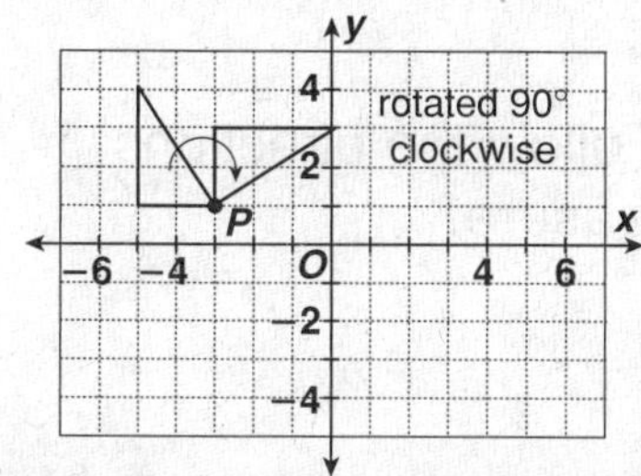

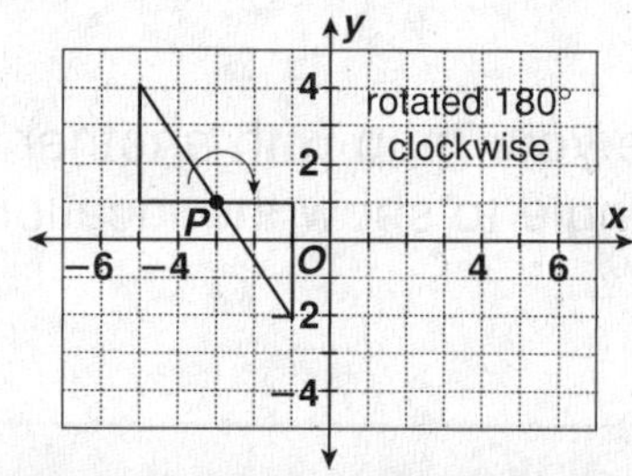

7. When you rotate a figure about a point, does the point of rotation move? __________

- You can reflect a figure across a line.

△*JKL* is reflected across the *x*-axis.

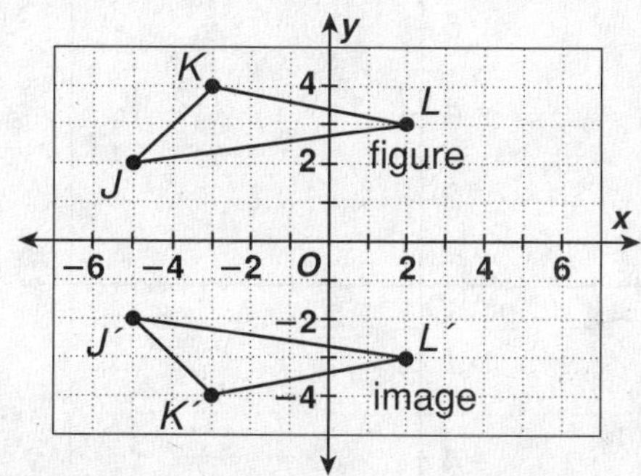

△*PQR* is reflected across the *y*-axis.

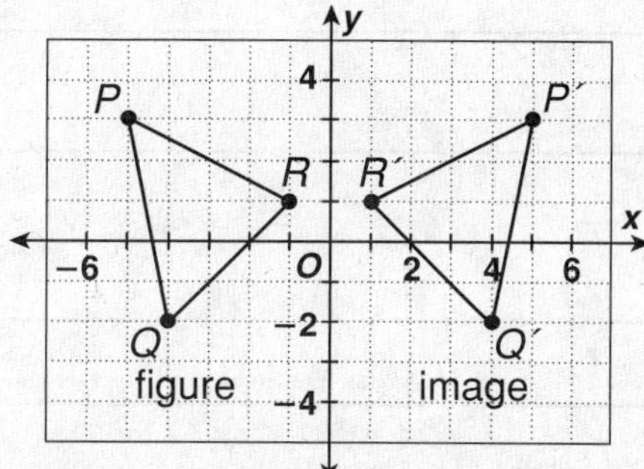

8. When you reflect a figure across the *x*-axis, do the *x*-coordinates change or remain the same? __________

9. When you reflect a figure across the *x*-axis, do the *y*-coordinates change or remain the same? __________

10. When you reflect a figure across the *y*-axis, do the *x*-coordinates change or remain the same? __________

11. When you reflect a figure across the *y*-axis, do the *y*-coordinates change or remain the same? __________

Name ______________________ Date __________ Class __________

LESSON 8-10

Challenge

Follow That Transformation

You alone know the location of the secret treasure. You want to draw a map to direct your friend to the treasure, but you must do so in code. Use transformations to make the code.

First, draw a triangle on the coordinate plane below.

Next, list in order 5 transformations that must be done to the original figure. The treasure is located in the image of the last transformed triangle.

Then exchange your map with another student. Follow the directions and draw a triangle to show the location of the treasure.

1. ______________________________

2. ______________________________

3. ______________________________

4. ______________________________

5. ______________________________

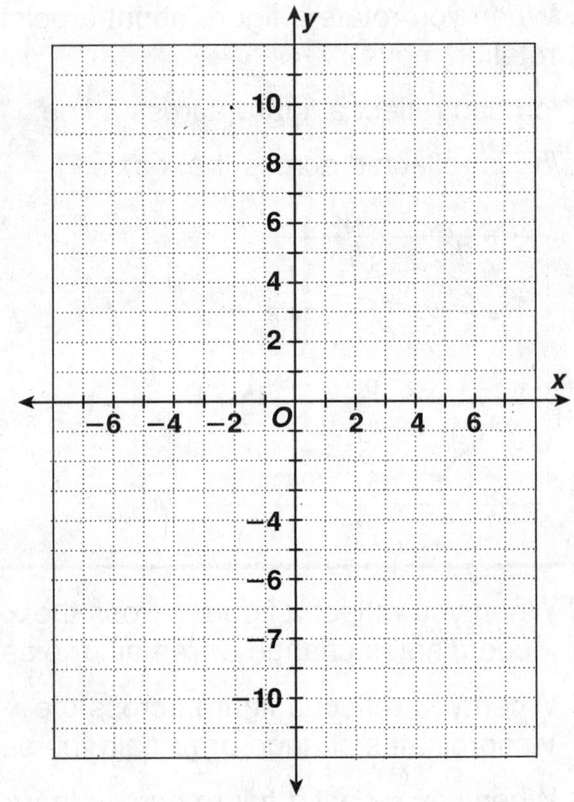

Name ______________________ Date __________ Class __________

LESSON **8-10**

Problem Solving

Translations, Reflections, and Rotations

Write the correct answer.

Clock 1

Clock 2

1. If you reflect the hands of clock 1 across a line from 12 to 6, what time will it show?

____

2. If you rotate the hour hand on clock 2 by 90° clockwise, what time will it be?

____

3. The hands on clock 1 show 7:00 after a transformation of one hand. What was the transformation?

____

____

4. The hands on clock 2 show 9:00 after a transformation. Name 2 different transformations that could produce this change.

____

____

____

Choose the letter for the best answer.

5. What transformation of triangle 1 created triangle 2?

A translation 3 units right and 1 unit down

B translation 8 units right and 1 unit down

C rotation of 180° about the origin

D reflection across the *y*-axis

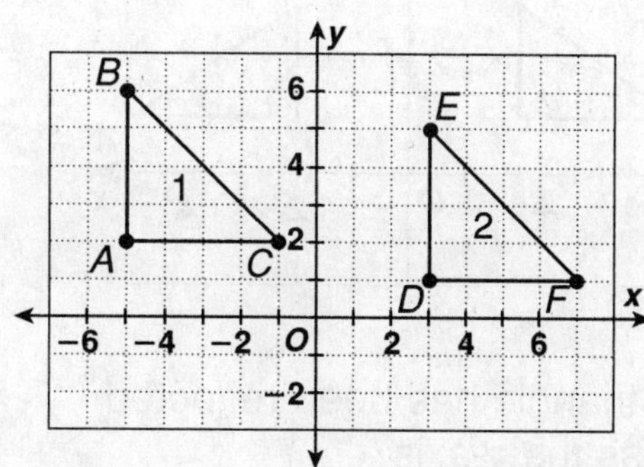

6. If you rotate triangle 2 90° clockwise about vertex *D*, what will be the coordinates of the new triangle?

F *D*′(3, 1), *E*′(7, 1), *F*′(3, −3)

G *D*′(3, 1), *E*′(3, −3), *F*′(7, 1)

H *D*′(3, 1), *E*′(−4, 1), *F*′(−3, 3)

J *D*′(3, 1), *E*′(−3, 3), *F*′(−7, 1)

7. If you reflect triangle 1 across the *x*-axis, what will be the coordinates of the new triangle?

A *A*′(5, 2), *B*′(5, 6), *C*′(1, 2)

B *A*′(−5, 0), *B*′(-5, −4), *C*′(−1, 0)

C *A*′(5, −2), *B*′(5, −6), *C*′(1, −2)

D *A*′(−5, −2), *B*′(−5, −6), *C*′(−1, −2)

Name ______________________ Date __________ Class __________

LESSON 8-10

Reading Strategies
Use Graphics

A **transformation** moves a figure but does not change its shape. The figure that moves is called the **image** of the original figure.

Here are three kinds of transformations.

- **Translation**
  Slide a figure to a new position.

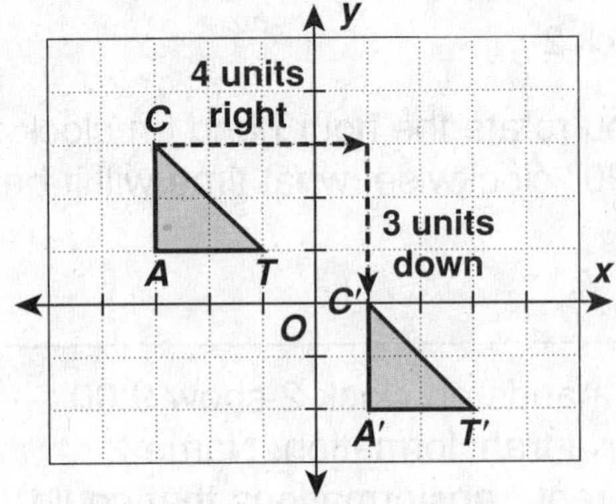

△*CAT* has moved four units to the right and three units down.

- **Reflection**
  Flip a figure across the *y*-axis.

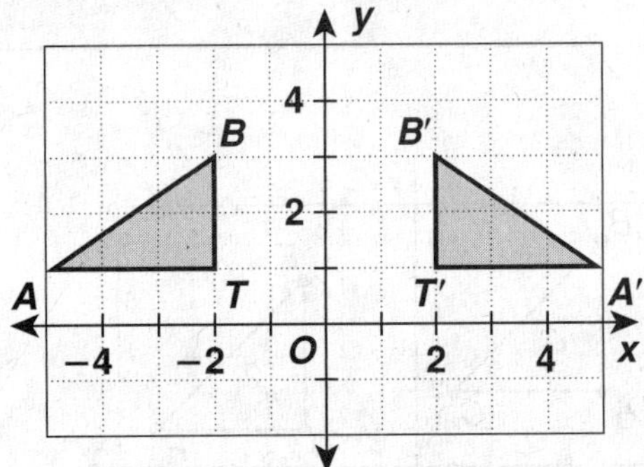

The triangle has been reflected across the *y*-axis.

- **Rotation**
  Turn a figure around a point.

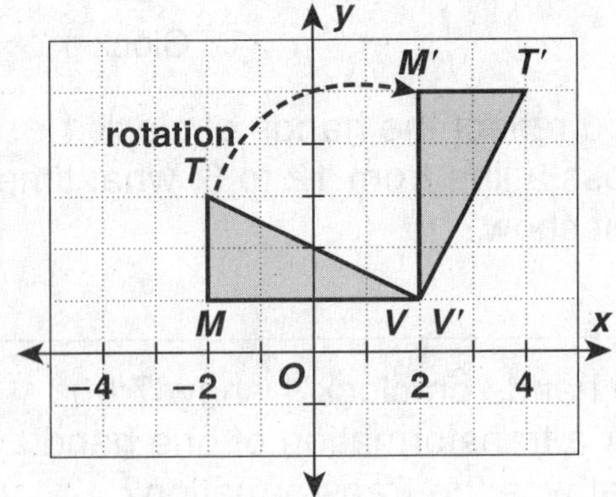

△*MTV* has been rotated 90° clockwise around vertex *V*.

1. Translate △DEF six units to the right and two units down.

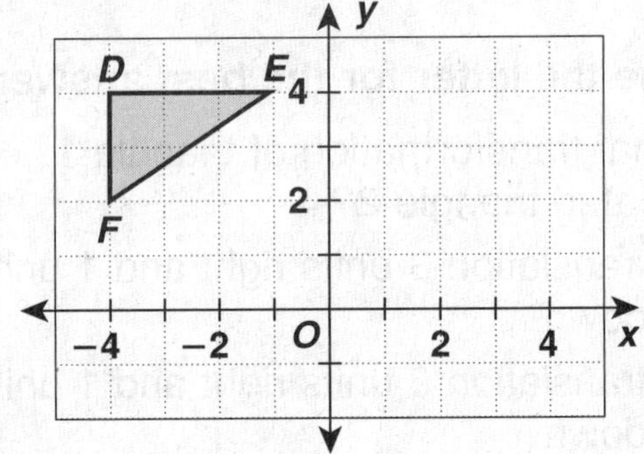

2. Reflect the figure across the *y*-axis.

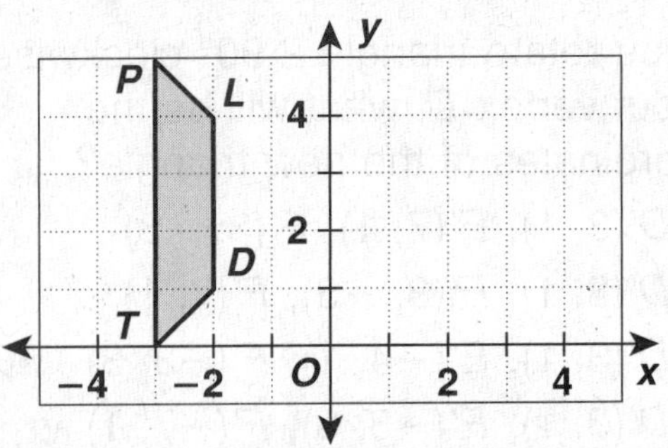

Name ______________________ Date __________ Class __________

LESSON 8-10

Puzzles, Twisters & Teasers

Transformer!

Across

1. A way to produce a new image from a given one

6. Not moving

8. A reflection is a _____ image

9. _____ of reflection

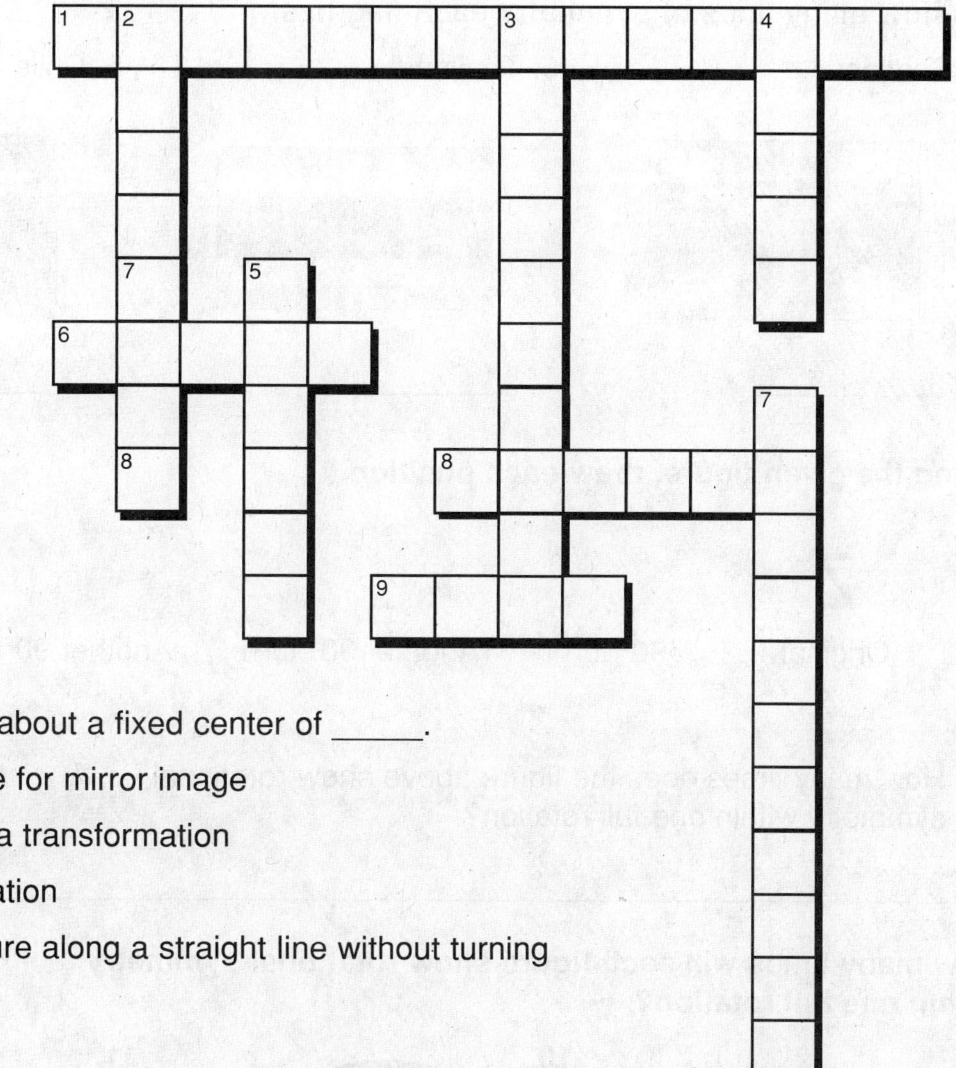

Down

2. A figure turns about a fixed center of _____.

3. Another name for mirror image

4. The result of a transformation

5. _____ of rotation

7. To slide a figure along a straight line without turning

Name ______________________ Date __________ Class __________

LESSON 8-11

Practice A

Symmetry

Draw all the lines of symmetry in each figure.

1.

2.

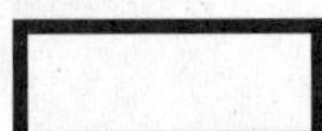

3.

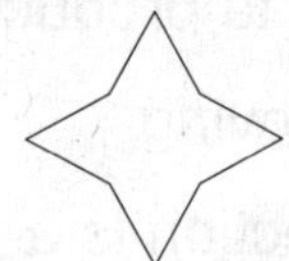

Tell how many lines of symmetry each flag has.

4. Scotland

____

5. Thailand

____

6. Chile

____

Using the given figure, draw each position.

7.

Z

Original A 90° turn Another 90° turn Another 90° turn

8. How many times does the figure above show rotational symmetry within one full rotation?

____

How many times will each figure show rotational symmetry within one full rotation?

9.

____

10.

____

11.

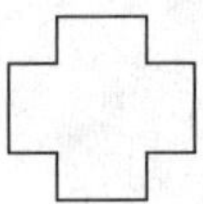

____

Name ______________________ Date __________ Class __________

LESSON **8-11**

Practice B

Symmetry

Decide whether each figure has line symmetry. If it does, draw all the lines of symmetry.

1.

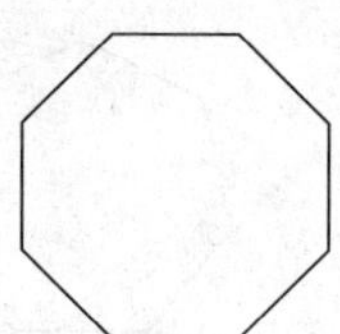

2.

3.

Find and draw all the lines of symmetry in each flag.

4. Iceland

5. Nauru

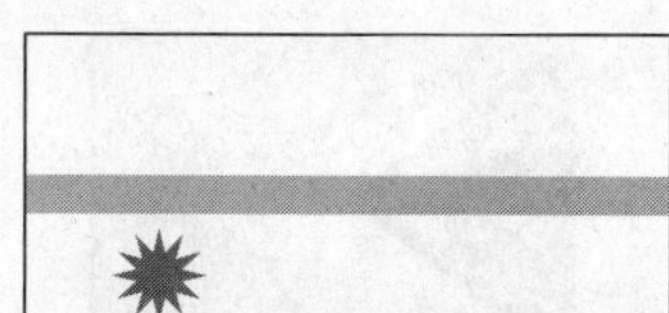

6. Burundi

Tell how many times each figure will show rotational symmetry within one full rotation.

7.

____

8.

____

9.

____

10.

____

11.

____

12.

____

Name ______________________ Date ____________ Class ____________

LESSON 8-11

Practice C
Symmetry

Decide whether each figure has line symmetry. If it does, draw all the lines of symmetry.

1.

2.

3.

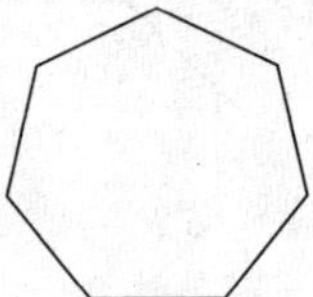

Find and draw all the lines of symmetry in each flag.

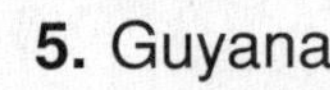

4. Great Britain

5. Guyana

6. Trinidad and Tobago

Tell how many times each figure will show rotational symmetry within one full rotation.

7.

____

8.

____

9.

____

Describe the smallest angle of rotational symmetry for each figure.

10.

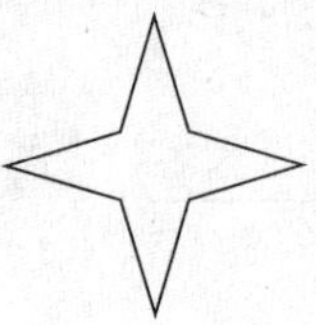

____

11.

____

12 .

____

Name ______________________ Date __________ Class __________

LESSON **8-11**

Reteach

Symmetry

A figure has **line symmetry** if it can be folded along a line so that the two halves match exactly. The fold line of a figure is called a **line of symmetry**. A figure can have more than one line of symmetry.

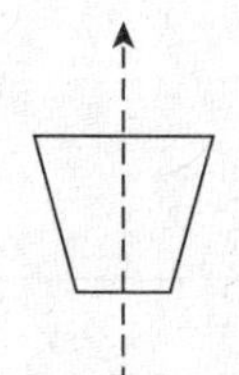

1 line of symmetry

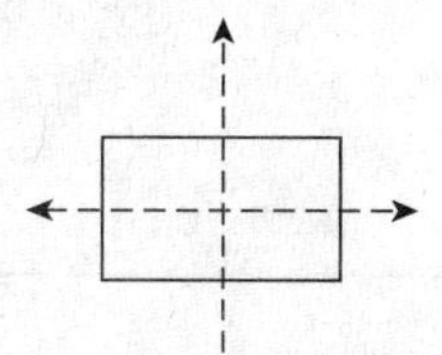

2 lines of symmetry

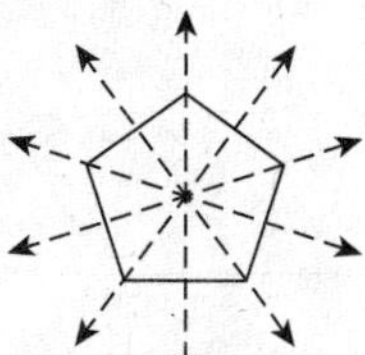

5 lines of symmetry

Trace each figure and cut it out. Fold the figure on each line of symmetry. Tell how many lines of symmetry each figure has.

1.

____

2.

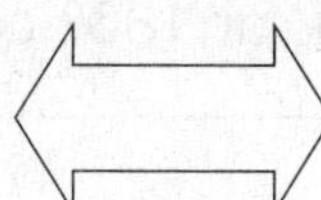

____

3.

____

A figure has **rotational symmetry** if it can be turned less than a full turn (360°) and look exactly like the original figure.

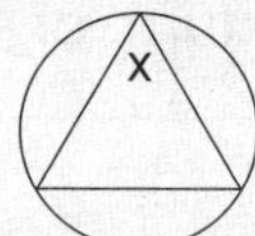

Original figure

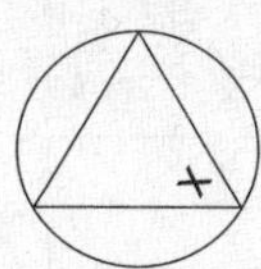

Rotation 1

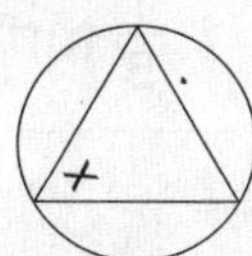

Rotation 2

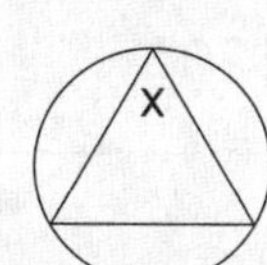

Rotation 3 – same as original figure

Trace each figure. Mark an X at the top of the tracing. Put the tracing on top of the figure and rotate it until it matches. Tell the number of rotations until the X is at the top again.

4.

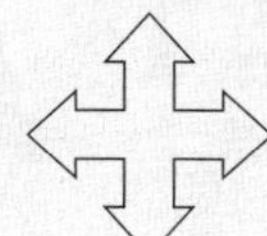

5.

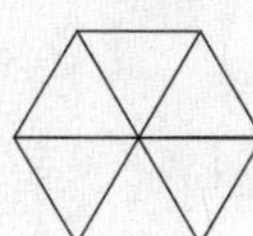

6.

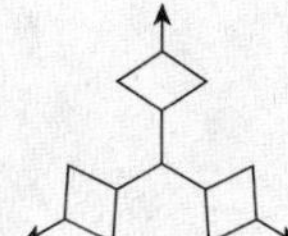

Name ______________________ Date ____________ Class ____________

LESSON 8-11

Challenge

Design Your Own Quilt

This quilt will be a collection of individual squares. In each square, draw a different figure that has the given rotational symmetry.

For example, 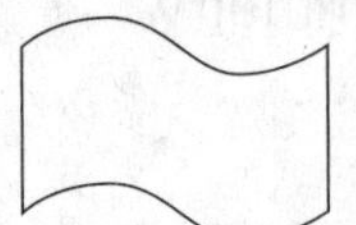 has $\frac{1}{2}$-turn, or 180°, symmetry.

$\frac{1}{4}$ turn, or 90°	$\frac{1}{2}$ turn, or 180°	$\frac{1}{4}$ turn, or 90°
$\frac{1}{3}$ turn, or 120°	$\frac{1}{5}$ turn, or 72°	$\frac{1}{3}$ turn, or 120°
$\frac{1}{4}$ turn, or 90°	$\frac{1}{2}$ turn, or 180°	$\frac{1}{4}$ turn, or 90°
$\frac{1}{3}$ turn, or 120°	$\frac{1}{6}$ turn, or 60°	$\frac{1}{3}$ turn, or 120°

Name ______________________ Date ____________ Class ____________

LESSON 8-11

Problem Solving
Symmetry

Write the correct answer.

Use the logo of an Australian TV station for Exercises 1 and 2.

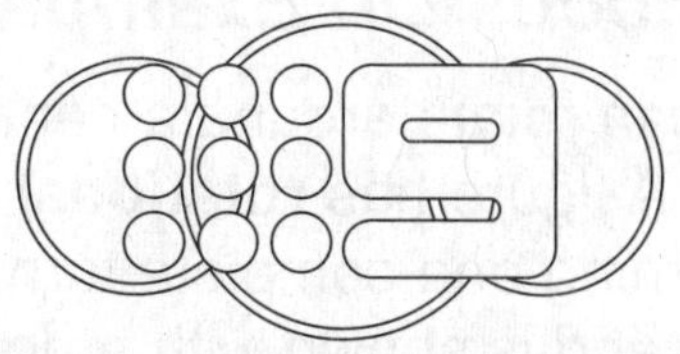

1. How many lines of symmetry do the dots in the Channel 9 logo have?

____

____

2. How many times will the dots show rotational symmetry in 1 rotation of 360°? What is the smallest angle of rotational symmetry for the dots?

____

3. The figure on an emblem will show rotational symmetry 8 times within a single rotation. What is the smallest angle of rotational symmetry for the emblem?

____

4. Draw a figure that shows rotational symmetry 5 times within a 360° rotation.

Choose the letter for the best answer.

5. Which is a true statement about the stained glass window?

   A It is asymmetrical.

   B It has 1 line of symmetry.

   C It will show rotational symmetry 4 times within a full rotation.

   D It has 2 lines of symmetry.

6. Which is a true statement about the time shown on the clocks?

3:08	4:00	6:13	1:30
1	2	3	4

   F Only clocks 1 and 2 have a line of symmetry.

   G All of the clocks have at least 1 line of symmetry.

   H All of the clocks are asymmetrical.

   J Only clocks 1 and 4 have a line of symmetry.

Name ______________________________ Date ____________ Class ____________

LESSON 8-11

Reading Strategies

Focus On Example/Non-Example

A figure can rotate around a central point called the **center of rotation.** A figure has **rotational symmetry** if it rotates onto itself before turning one complete turn, or 360°. The center of rotation is a point and is shown with a dot.

If you rotate the letter Z around the central point, it will rotate onto itself halfway around the circle, or at 180°. This figure *has* rotational symmetry

Rotational Symmetry

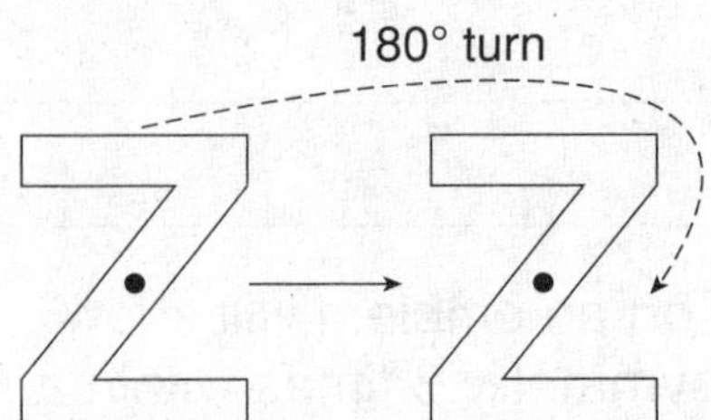

If you rotate the letter E around the central point, it will not rotate onto itself until it has made one full rotation around the circle, at 360°.
The figure *does not have* rotational symmetry.

No Rotational Symmetry

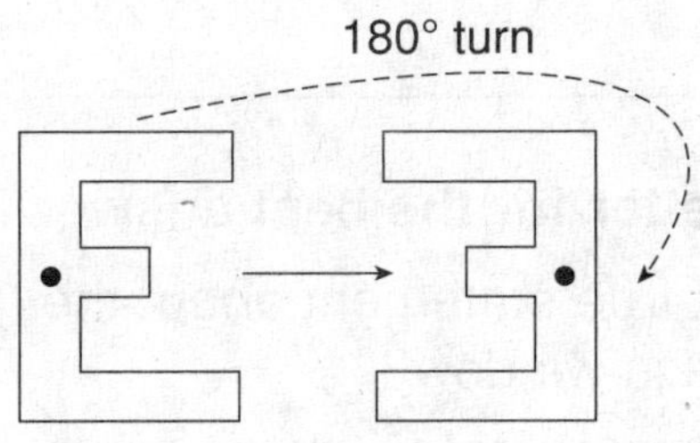

Trace and cut out a copy of each figure. Rotate the cut figure over the same figure on the page. Do these figures have rotational symmetry? Write yes or no.

1. 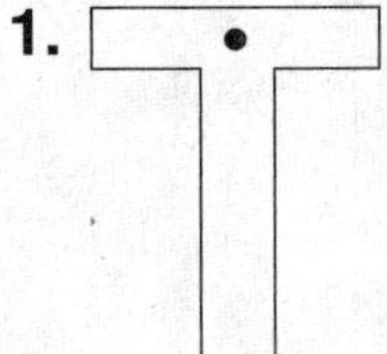

____

2. 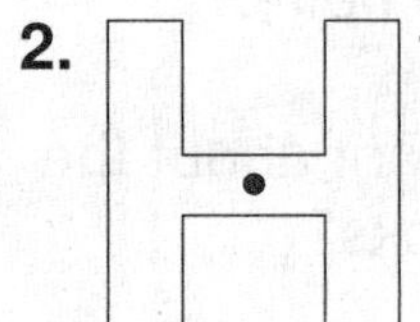

____

3. 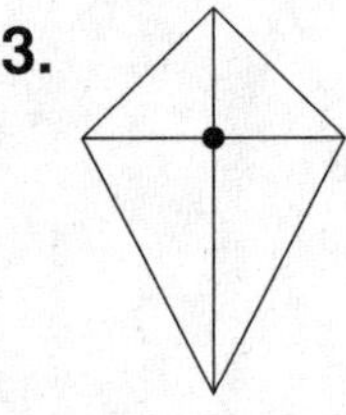

____

4. 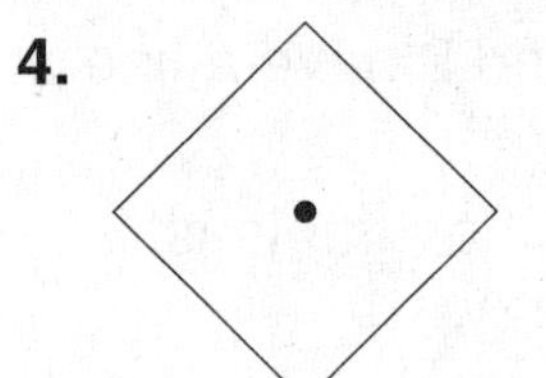

____

Name ______________________ Date ___________ Class ___________

LESSON 8-11

Puzzles, Twisters & Teasers

This Is Hairy!

Determine the answer for each. Use the letters of your answer to solve the riddle.

1. How many lines of symmetry does this figure have?

A 2 **B** 3 **C** 4 **D** None

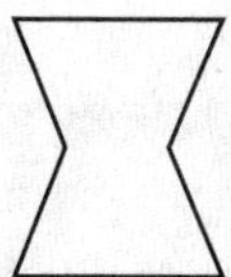

2. How many lines of symmetry does this figure have?

O 1 **A** 2 **E** 3 **H** None

3. How many lines of symmetry does this figure have?

T 2 **L** 6 **M** 16 **R** None

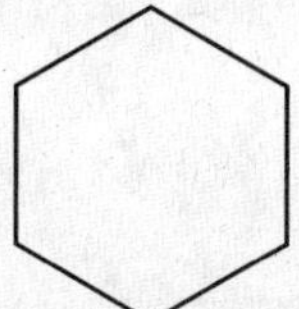

4. How many lines of symmetry does this figure have?

B 2 **S** 3 **T** 4 **H** None

5. How many lines of symmetry does this figure have?

T 2 **S** 3 **T** 4 **P** None

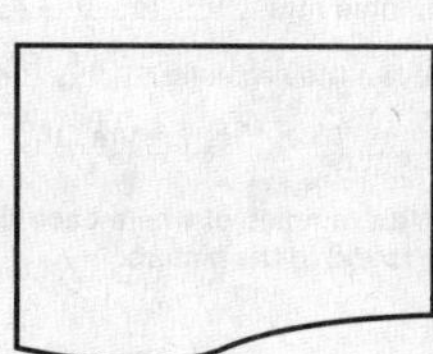

What kind of portable computer does a rabbit like to use?

___ 1   ___ 3   ___ 2   ___ 5   ___ 4   ___ 2   ___ 5

LESSON 8-1 Practice A
Building Blocks of Geometry

Write the following in geometric notation.

1. line *EF* $\overleftrightarrow{EF}$
2. ray *RS* $\overrightarrow{RS}$
3. line segment *JK* $\overline{JK}$

Choose the letter for the best answer.

4. Identify a line.
   A $\overline{BD}$ C $\overleftrightarrow{CB}$
   B $\overleftrightarrow{AD}$ (D) $\overleftrightarrow{BD}$

5. Identify a ray.
   (F) $\overrightarrow{AC}$ H $\overleftrightarrow{CD}$
   G $\overline{AD}$ J $\overrightarrow{CB}$

6. Identify a line segment.
   A $\overrightarrow{DB}$ C $\overline{CB}$
   (B) $\overline{DB}$ D $\overline{AD}$

7. Identify a plane.
   F plane *AB* (H) plane *ABD*
   G plane *ADE* J plane *BCF*

Identify the figures in the diagram. Possible Answers:

8. two lines $\overleftrightarrow{XY}$, $\overleftrightarrow{XZ}$, $\overleftrightarrow{YZ}$
9. three rays $\overrightarrow{XY}$, $\overrightarrow{YX}$, $\overrightarrow{XZ}$, $\overrightarrow{ZX}$, $\overrightarrow{YZ}$, $\overrightarrow{ZY}$
10. three points *X, Y, Z*
11. three line segments $\overline{XY}$, $\overline{XZ}$, $\overline{YZ}$
12. a plane plane *XYZ*

Use the figure for Exercise 13.

13. Identify which line segments are congruent.
    $\overline{AB}$ and $\overline{CD}$; $\overline{AC}$ and $\overline{BD}$

LESSON 8-1 Practice B
Building Blocks of Geometry

Identify the figures in the diagram. Possible answers:

1. three points *A, B, C, D*
2. one line $\overleftrightarrow{BC}$
3. a plane plane *ABC*
4. four rays $\overrightarrow{BD}$, $\overrightarrow{CD}$, $\overrightarrow{CB}$, $\overrightarrow{BC}$
5. three line segments $\overline{AB}$, $\overline{AC}$, $\overline{BD}$, $\overline{CD}$, $\overline{BC}$

Identify the figures in the diagram. Possible answers:

6. four points *M, N, O, P*
7. three lines $\overleftrightarrow{MN}$, $\overleftrightarrow{MO}$, $\overleftrightarrow{NO}$
8. a plane plane *MNO*
9. three rays $\overrightarrow{MO}$, $\overrightarrow{MN}$, $\overrightarrow{NM}$, $\overrightarrow{NO}$, $\overrightarrow{OM}$, $\overrightarrow{ON}$, $\overrightarrow{OP}$
10. four line segments $\overline{MN}$, $\overline{MO}$, $\overline{NO}$, $\overline{OP}$

Identify the figures in the diagram. Possible answers:

11. four points *Q, R, S, T*
12. two lines $\overleftrightarrow{QS}$, $\overleftrightarrow{RT}$
13. a plane plane *QRS*
14. four rays $\overrightarrow{QS}$, $\overrightarrow{SQ}$, $\overrightarrow{RT}$, $\overrightarrow{TR}$
15. five line segments $\overline{QR}$, $\overline{RS}$, $\overline{QS}$, $\overline{ST}$, $\overline{RT}$
16. Identify the line segments that are congruent in the figure.
    $\overline{AB}$ and $\overline{DE}$; $\overline{BC}$ and $\overline{EF}$;
    $\overline{AD}$, $\overline{BE}$, and $\overline{CF}$

LESSON 8-1 Practice C
Building Blocks of Geometry

Identify the figures in the diagram.

1. six points *A, B, C, D, E, F*
2. a line $\overleftrightarrow{BE}$
3. planes (list 4 possible names) Possible answers:
   plane *ACE*, plane *BCD*,
   plane *ABD*, plane *CEF*
4. three rays $\overrightarrow{CA}$, $\overrightarrow{DF}$, $\overrightarrow{BD}$, $\overrightarrow{EC}$, $\overrightarrow{BE}$, $\overrightarrow{EB}$, $\overrightarrow{BA}$, $\overrightarrow{EF}$
5. five line segments $\overline{AB}$, $\overline{BC}$, $\overline{AD}$, $\overline{DB}$, $\overline{BE}$, $\overline{EC}$,
   $\overline{CF}$, $\overline{DE}$, $\overline{EF}$, $\overline{AC}$, $\overline{DF}$

Give examples of where each figure could be found in the picture.

6. points
   Possible answer: any corner of the house or of a window or door
7. planes
   Possible answer: the ground
8. rays
   Possible answer: the sunlight
9. line segments
   Possible answer: all of the line segments that make up the house
10. congruent line segments
    Possible answer: opposite sides of windows and door

LESSON 8-1 Reteach
Building Blocks of Geometry

You can think of real-life objects to represent geometric terms.

Point: a polka dot	Line: a straight road in both directions	Ray: flashlight beam
Point *A*	Line *BC*, or $\overleftrightarrow{BC}$	Ray *PQ*, or $\overrightarrow{PQ}$
Line segment: a ruler	Part of a plane: table top	Congruent line segments: lines on an index card
Line segment *GH*, or $\overline{GH}$	Plane *DEF*	$\overline{WX} \cong \overline{YZ}$

For 1–5, use geometry notation to identify the figures.

1. three points
   J, K, L, M, N
2. three lines
   $\overleftrightarrow{JK}$, $\overleftrightarrow{KL}$, $\overleftrightarrow{JL}$
3. a plane
   Possible answer: plane *JMK*
4. three rays
   $\overrightarrow{LN}$, $\overrightarrow{JM}$, $\overrightarrow{JK}$, $\overrightarrow{KJ}$, $\overrightarrow{KL}$, $\overrightarrow{LK}$, $\overrightarrow{JL}$, $\overrightarrow{LJ}$
5. three line segments
   $\overline{JK}$, $\overline{KL}$, $\overline{JL}$, $\overline{JM}$, $\overline{LN}$
6. Identify the congruent line segments in the figure at the right.
   $\overline{AE} \cong \overline{AC}$; $\overline{CD} \cong \overline{DE}$

LESSON 8-1
Challenge
Plane Drawings

You can use a drawing to represent a part of a plane.

The horizontal planes below do not intersect. They are *parallel*. The dashed lines show that one plane lies behind the other plane.

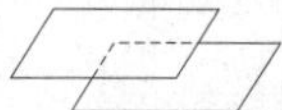

The steps below demonstrate how to sketch two intersecting planes.

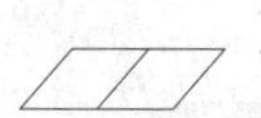

Draw a parallelogram.
Draw a line segment through its center.

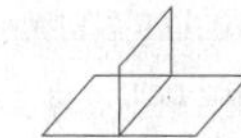

Use the segment as one side of a new parallelogram.

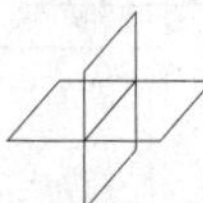

Draw the bottom part of a new plane.

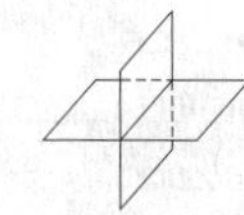

Dash the invisible edges.

Make a sketch of the planes described below.

1. Draw a pair of vertical planes that do not intersect. **Answers will vary. Possible answer given.**

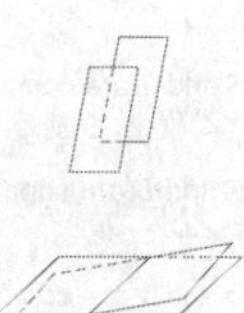

2. Draw a pair of planes that intersect at an angle different from the example sketched above. **Answers will vary. Possible answer given.**

LESSON 8-1
Problem Solving
Building Blocks of Geometry

Write the correct answer. **Answers may vary. Possible answers are given.**

The drawing shows a section of the Golden Gate Bridge in San Francisco.

1. Identify two lines that are suggested by the bridge.

   $\overleftrightarrow{JK}$ and $\overleftrightarrow{LM}$

2. Identify a ray and a line segment that are suggested by the bridge.

   $\overrightarrow{RQ}$ and $\overline{ST}$

3. Identify two lines in the figure that are in the same plane.

   $\overleftrightarrow{JK}$ and $\overleftrightarrow{LM}$

4. Identify a plane in the figure.

   plane *RLM*

Choose the letter for the best answer.

The drawing is an artist's sketch for an abstract painting.

5. Which line segment is congruent to $\overline{BC}$?

   A $\overline{AB}$ C $\overline{AC}$
   (B) $\overline{KJ}$ D $\overline{HM}$

6. Which line segment is congruent to $\overline{GM}$?

   F $\overline{AC}$ H $\overline{DE}$
   G $\overline{DF}$ **(J)** $\overline{AB}$

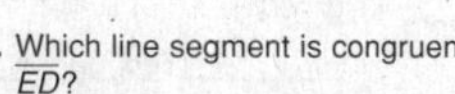

7. Which line segment is congruent to $\overline{ED}$?

   (A) $\overline{AC}$ C $\overline{DF}$
   B $\overline{CD}$ D $\overline{GH}$

8. Which line segment is congruent to $\overline{DF}$?

   F $\overline{ED}$ H $\overline{CJ}$
   (G) $\overline{HM}$ J $\overline{FM}$

LESSON 8-1
Reading Strategies
Vocabulary Development

A **point** marks an exact location in space. It is usually shown as a dot.

J Read: point *J*

A **line** is a straight path that extends forever in both directions.

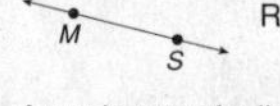

Read: line *MS* or $\overleftrightarrow{MS}$ or line *SM* or $\overleftrightarrow{SM}$

A **ray** is part of a line that has one endpoint and goes on forever in one direction.

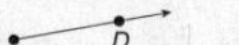

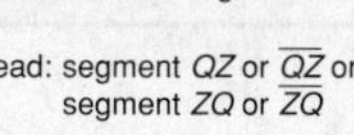

Read: ray *TD* or $\overrightarrow{TD}$

A **line segment** is part of a line between two endpoints. The two endpoints are used to name the line segment.

Read: segment *QZ* or $\overline{QZ}$ or segment *ZQ* or $\overline{ZQ}$

Congruent line segments have the same length. Tick marks are used to show congruent segments.

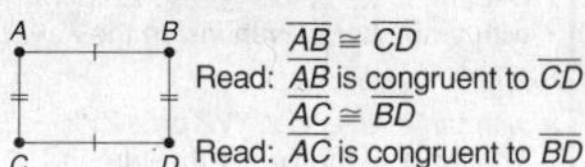

$\overline{AB} \cong \overline{CD}$
Read: $\overline{AB}$ is congruent to $\overline{CD}$
$\overline{AC} \cong \overline{BD}$
Read: $\overline{AC}$ is congruent to $\overline{BD}$

A flat surface that extends without end in all directions is called a **plane.** A plane is named by three points on the plane that are not on a line.

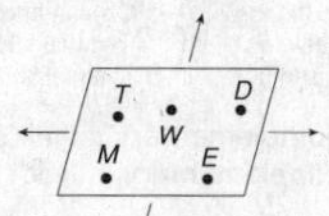

This plane could be named plane *MTW*.

Use the information above to answer each question.

1. What name is given to a path that goes on in both directions?

   line

2. What name is given to part of a line with two endpoints?

   line segment

3. What name is given to part of a line with only one endpoint?

   ray

4. What name is given to line segments with the same length?

   congruent line segments

5. How would you describe a plane?

   Possible answer: a flat surface that extends without end in all directions

LESSON 8-1
Puzzles, Twisters & Teasers
Are You Barking Up the Wrong Tree?

Find and circle words from the list in the word search (horizontally, vertically or diagonally). Then find a word in the word search that solves the riddle. Circle it and write it on the line.

points	notation	ray	line	segment
plane	congruent	figure	geometric	endpoint

```
G E O M E T R I C C E R
Z N M H T G A Y O V G Y
C D E T U I Y P N B H U
V P L A N E Z A G N I S
B O I B I P L M R J O E
N I V A K F I G U R E G
M N O R B H U I E Q W M
L T M K W E L I N E V E
K I J G N O T A T I O N
L D V G P O I N T S L T
```

How can you identify a dogwood tree?

By its **B A R K**

LESSON 8-2
Practice A
Classifying Angles

Tell whether each angle is acute, right, obtuse, or straight.

1. 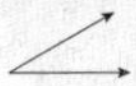 acute
2.  obtuse
3. 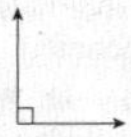 right

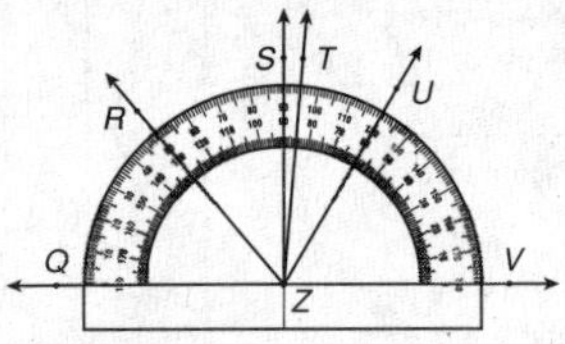

Use the diagram to complete Exercises 4-8.

4. Add the measures of ∠UZV and ∠TZU. $60° + 25° = 85°$
5. Are ∠UZV and ∠TZU complementary, supplementary, or neither? neither
6. Add the measures of ∠QZR and ∠RZS. $50° + 40° = 90°$
7. Are ∠QZR and ∠RZS complementary, supplementary, or neither? complementary
8. Are ∠QZS and ∠VZS complementary, supplementary, or neither? supplementary
9. Angles G and H are supplementary. If m∠G is 80°, what is m∠H? 100°
10. Angles M and N are complementary. If m∠M is 20°, what is m∠N? 70°

LESSON 8-2
Practice B
Classifying Angles

Tell whether each angle is acute, right, obtuse, or straight.

1.  straight
2.  acute
3.  obtuse

Use the diagram to tell whether the angles are complementary, supplementary or neither.

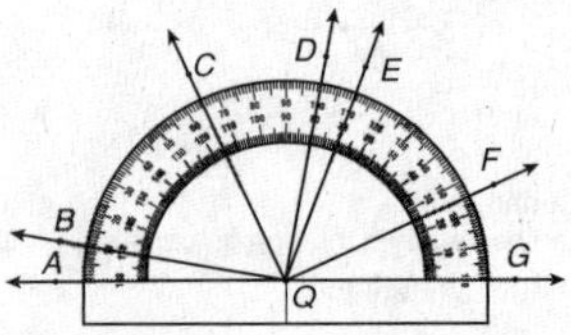

4. ∠AQC and ∠GQC supplementary
5. ∠BQD and ∠DQE neither
6. ∠CQE and ∠EQF complementary
7. ∠GQF and ∠FQE neither
8. ∠BQC and ∠DQC complementary
9. Angles W and X are supplementary. If m∠W is 37°, what is m∠X? 143°
10. Angles S and T are complementary. If m∠S is 64°, what is m∠T? 26°
11. Angles C and D are supplementary. If m∠C is 83°, what is m∠D? 97°
12. Angles U and V are complementary. If m∠U is 41°, what is m∠V? 49°

LESSON 8-2
Practice C
Classifying Angles

Tell whether each angle is acute, right, obtuse, or straight.

1.  right
2.  obtuse
3. 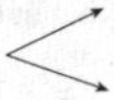 acute

Use the diagram to tell whether the angles are complementary, supplementary or neither.

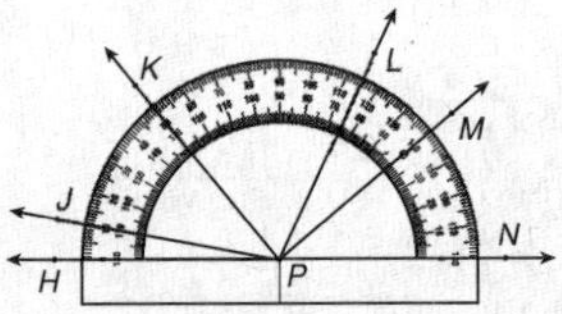

4. ∠JPK and ∠KPL neither
5. ∠LPK and ∠MPL complementary
6. ∠HPM and ∠MPN supplementary
7. ∠LPM and ∠KPJ neither
8. Angles L and M are complementary. If m∠L is 16°, what is m∠M? 74°
9. Angles R and S are supplementary. If m∠R is 78°, what is m∠S? 102°

Classify each pair of angles as complementary or supplementary. Then find the missing angle measure.

10. 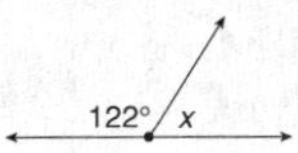 122°, x — supplementary; 58°
11. 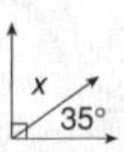 x, 35° — complementary; 55°

LESSON 8-2
Reteach
Classifying Angles

Two rays with a common endpoint form an **angle.** The common endpoint is called the **vertex.** You can name an angle three ways:

- with the letter at the vertex: ∠Y.
- with a number written inside the angle: ∠1.
- with three letters: ∠XYZ or ∠ZYX. The letter of the vertex must be in the middle.

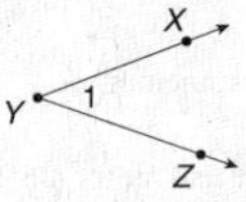

Angles are measured in degrees and classified by their measures.

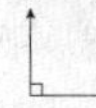

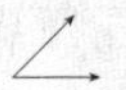

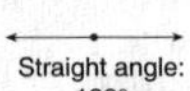

Right angle: 90° — Acute angle: between 0° and 90° — Obtuse angle: between 90° and 180° — Straight angle: 180°

Two angles are **complementary** if their sum is 90°.
Two angles are **supplementary** if their sum is 180°.

Use the figure to complete the statements.

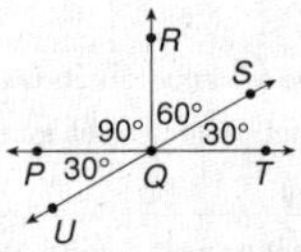

1. ∠PQR measures 90°. It is a right angle.
2. ∠RQS measures 60°. It is an acute angle.
3. ∠UQR measures 30° + 90°, or 120°. It is an obtuse angle.
4. ∠PQT measures 90° + 60° + 30°, or 180°. It is a straight angle.

Use the figure to complete the statements.

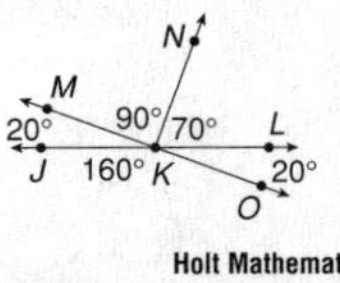

5. ∠NKL and ∠LKO are complementary angles since 70° + 20° = 90°.
6. ∠JKO and ∠OKL are supplementary angles since 160° + 20° = 180°.

LESSON 8-2

Challenge

Angling Clocks

The hands of a clock form an angle. At 9 o'clock, the hour hand points to 9, and the minute hand points to 12. The angle formed measures 90°.

But how do you measure the angle when both hands are between numbers?

- There are 5 minutes between two consecutive numbers on a clock.
- Each time the minute hand moves 12 minutes, or $\frac{1}{5}$ the distance around the clock, the hour hand moves $\frac{1}{5}$ the distance between two consecutive numbers.

What is the angle between the hands at 7:24?

When the minute hand moves 24 minutes, the hour hand moves 2 sections. So, there are $(7 \cdot 5 + 2) - 24$, or 13, sections between the hands. Each section is $\frac{1}{60}$ of an hour, and there are 360° on a clock face.

$13 \cdot \frac{1}{60} \cdot 360 = 13 \cdot 6 = 78$

The angle between the hands measures 78°.

1. Draw the clock hands and find the angle between them at 5 o'clock.

150°

2. Draw the clock hands and find the angle between them at 8 o'clock.

120°

3. Find the angle between the hands at 1 o'clock. 30°
4. Find the angle between the hands at 2 o'clock. 60°
5. Find the angle between the hands at 12:24. 132°
6. Find the angle between the hands at 3:48. 174°
7. Find the angle between the hands at 3:12. 24°
8. Find the angle between the hands at 1:36. 168°

LESSON 8-2

Problem Solving

Classifying Angles

Write the correct answer.

The drawing shows a scene on a calendar.

1. ∠1 and ∠2 are complementary angles. If ∠1 measures 35°, what is the measure of ∠2?

55°

2. ∠3 and ∠4 are supplementary angles. If ∠3 measures 50°, what is the measure of ∠4?

130°

3. Which angle is an obtuse angle: ∠6 or ∠7?

∠6

4. Which angle labeled on the drawing is a right angle?

∠5

Choose the letter for the correct answer.

Use the diagram to complete Exercises 5 and 6.

5. Which of the following could be the measures of ∠*TZU* and ∠*QZR*?
   A m∠*TZU* = 55° and m∠*QZR* = 55°
   B m∠*TZU* = 25° and m∠*QZR* = 95°
   C m∠*TZU* = 80° and m ∠*QZR* = 100°
   (D) m∠*TZU* = 35° and m∠*QZR* = 80°
6. If ∠*RZS* measures 35°, what is the measure of ∠*SZT*?
   F 155°
   (G) 145°
   H 55°
   J 45°
7. ∠*A* and ∠*B* are complementary angles. The measure of ∠*B* is 4 times the measure of ∠*A*. What are the measures of the angles?
   A m∠*A* = 16° and m∠*B* = 64°
   (B) m∠*A* = 18° and m∠*B* = 72°
   C m∠*A* = 36° and m ∠*B* = 144°
   D m∠*A* = 45° and m∠*B* = 135°
8. The hands of a clock form an acute angle at 1:00. What type of angle do they form at 4:00?
   F acute
   G right
   (H) obtuse
   J straight

LESSON 8-2

Reading Strategies

Use Manipulatives

You use a ruler to measure the length of a line. You use a **protractor** to measure the degrees of an angle.

Using a protractor, follow these steps to measure an angle.

place vertex of angle here

0 point

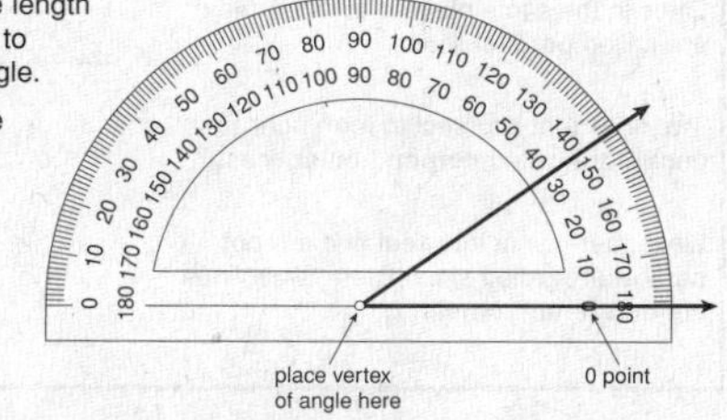

Step 1: Place the vertex of the angle under the hole in the middle of the protractor.

Step 2: Line up the 0 line on the protractor with one of the rays. This ray is lined up with the 0 line on the right side of the protractor, so you will use the inside numbers to find the measure of the angle.

Step 3: Follow the inside numbers from 0 until you reach the other ray. Notice that the other ray is on 35°. The measure of the angle is 35°.

Use the information above to answer each question.

1. What tool is used to measure an angle? a protractor
2. Where is the vertex of an angle placed on your protractor to measure the angle?

under the hole in the middle of the protractor

3. What part of an angle lines up with the 0 line on the protractor?

one of the two rays

Use a protractor to measure these angles.

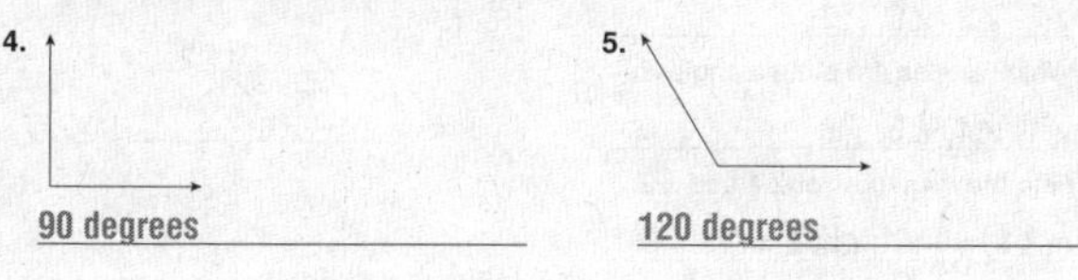

4. 90 degrees

5. 120 degrees

LESSON 8-2

Puzzles, Twisters & Teasers

Off the Hook!

Draw lines to match each term with the correct figure. Each figure has a corresponding letter. Use the letters to solve the riddle.

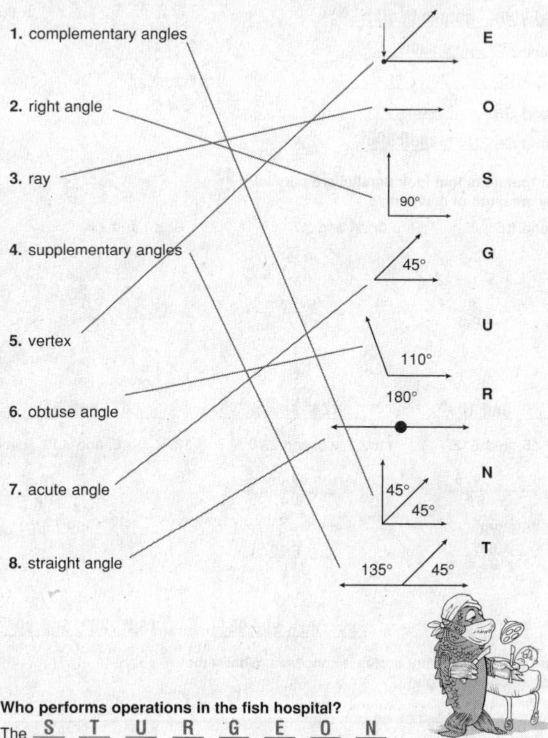

Who performs operations in the fish hospital?

The S T U R G E O N

LESSON 8-3
Practice A
Angle Relationships

1. Draw a pair of parallel lines.

2. Draw a pair of perpendicular lines.

Tell whether each statement is true or false.

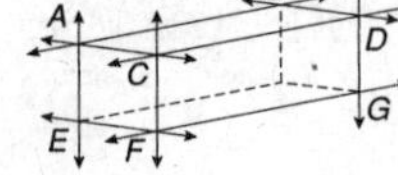

3. $\overleftrightarrow{AC}$ and $\overleftrightarrow{AB}$ appear parallel. **false**
4. $\overleftrightarrow{AC}$ and $\overleftrightarrow{EF}$ appear parallel. **true**
5. $\overleftrightarrow{AB}$ and $\overleftrightarrow{CD}$ appear perpendicular. **false**
6. $\overleftrightarrow{CF}$ and $\overleftrightarrow{FG}$ appear perpendicular. **true**
7. $\overleftrightarrow{EF}$ and $\overleftrightarrow{DG}$ appear to be skew. **true**
8. $\overleftrightarrow{CD}$ and $\overleftrightarrow{DG}$ appear perpendicular. **true**

Line *x* is parallel to line *y*. Find the measure of each angle.

9. ∠1 and ∠6

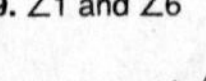

120° and 60°

10. ∠3 and ∠7

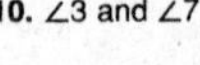

45° and 45°

11. ∠1 and ∠4

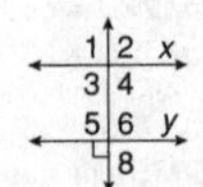

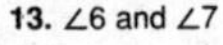

90° and 90°

12. ∠6 and ∠4

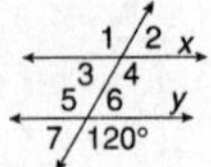

80° and 100°

13. ∠6 and ∠7

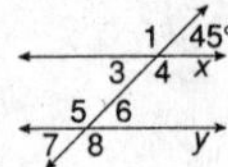

65° and 65°

14. ∠1 and ∠3

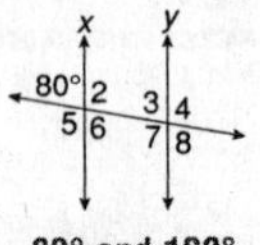

130° and 50°

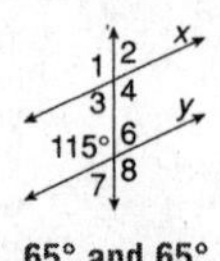

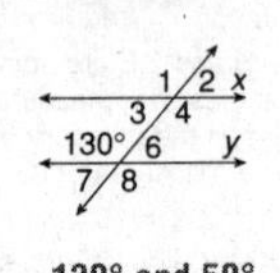

LESSON 8-3
Practice B
Angle Relationships

Tell whether the lines appear parallel, perpendicular, or skew.

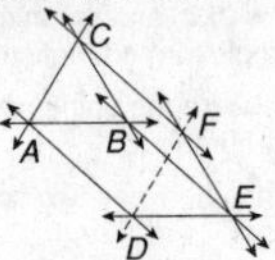

1. $\overleftrightarrow{AB}$ and $\overleftrightarrow{DE}$ **parallel**
2. $\overleftrightarrow{EF}$ and $\overleftrightarrow{CF}$ **perpendicular**
3. $\overleftrightarrow{AB}$ and $\overleftrightarrow{AD}$ **perpendicular**
4. $\overleftrightarrow{BC}$ and $\overleftrightarrow{DE}$ **skew**

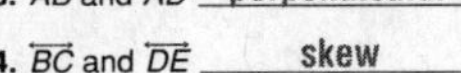

Tell whether the lines appear parallel, perpendicular, or skew.

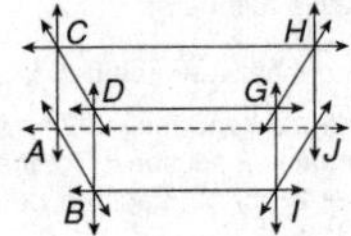

5. $\overleftrightarrow{BD}$ and $\overleftrightarrow{DG}$ **perpendicular**
6. $\overleftrightarrow{AB}$ and $\overleftrightarrow{BD}$ **perpendicular**
7. $\overleftrightarrow{DG}$ and $\overleftrightarrow{IJ}$ **skew**
8. $\overleftrightarrow{AB}$ and $\overleftrightarrow{CD}$ **parallel**

Line *x* ∥ line *y*. Find the measure of each.

9. ∠1 and ∠6

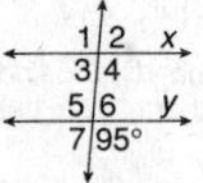

95° and 85°

10. ∠4 and ∠8

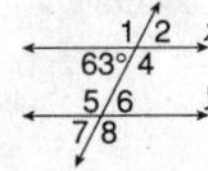

117° and 117°

11. ∠4 and ∠6

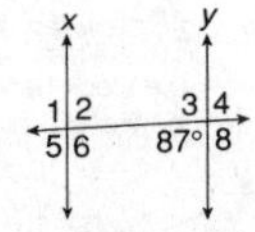

87° and 93°

12. ∠2 and ∠4

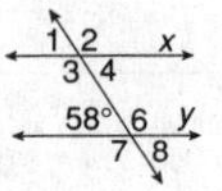

122° and 58°

13. ∠5 and ∠7

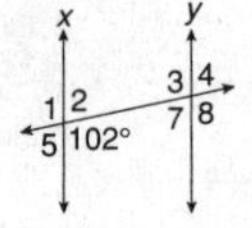

78° and 78°

14. ∠7 and ∠8

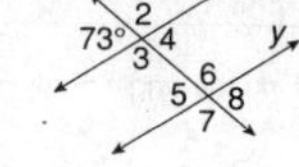

107° and 73°

LESSON 8-3
Practice C
Angle Relationships

Tell whether the lines appear parallel, perpendicular or skew.

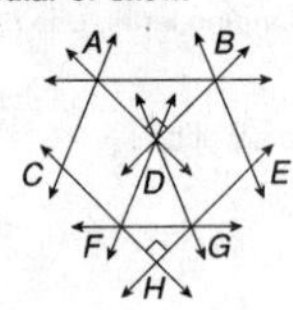

1. $\overleftrightarrow{DF}$ and $\overleftrightarrow{AC}$ **parallel**
2. $\overleftrightarrow{AD}$ and $\overleftrightarrow{BD}$ **perpendicular**
3. $\overleftrightarrow{AD}$ and $\overleftrightarrow{GE}$ **skew**
4. $\overleftrightarrow{DG}$ and $\overleftrightarrow{BE}$ **parallel**
5. $\overleftrightarrow{AB}$ and $\overleftrightarrow{GH}$ **skew**
6. $\overleftrightarrow{FH}$ and $\overleftrightarrow{GH}$ **perpendicular**

Assume that lines that look parallel are parallel. Find the measure of each angle.

7. ∠4 and ∠6

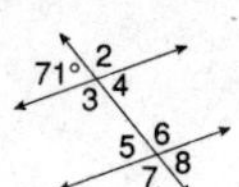

71° and 109°

8. ∠1 and ∠7

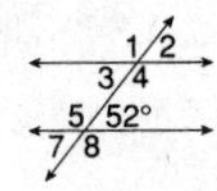

128° and 52°

9. ∠3 and ∠8

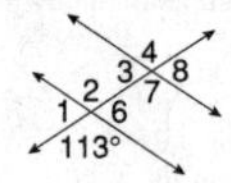

67° and 67°

10. ∠2, ∠5, and ∠12

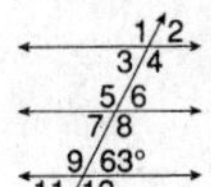

63°, 117°, and 117°

11. ∠1, ∠8, and ∠10

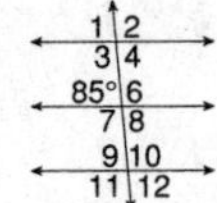

85°, 85°, and 95°

12. ∠3, ∠6, and ∠11

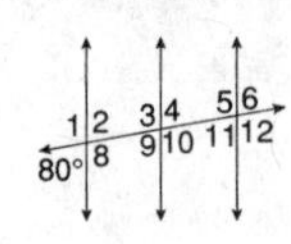

100°, 80°, and 80°

13. A pair of supplementary angles is congruent. What is the measure of each angle?

90°

LESSON 8-3
Reteach
Angle Relationships

Lines in the same plane that never meet are called **parallel** lines.

Two lines that intersect to form right angles are called **perpendicular** lines.

Lines that do not intersect and are not parallel are called **skew** lines. Skew lines are in different planes.

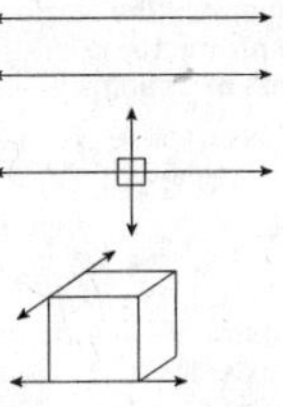

Use the figure to complete the statements.

1. $\overleftrightarrow{AB}$ and $\overleftrightarrow{DC}$ are **parallel** lines, since they never meet and lie in the same plane.
2. $\overleftrightarrow{DH}$ and $\overleftrightarrow{EH}$ are **perpendicular** lines, since they intersect to form right angles.
3. $\overleftrightarrow{AB}$ and $\overleftrightarrow{EH}$ are **skew** lines, since they do not intersect and are not parallel.

A **transversal** is a line that intersects two or more lines.

- When two parallel lines are intersected by a transversal, 8 angles are formed.
- The 4 acute angles are congruent, and the 4 obtuse angles are congruent.
- The sum of an acute angle and an obtuse angle is 180°.

Use the figure to answer the questions.

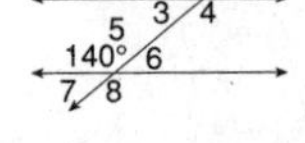

4. Which angles are acute angles?

∠2, ∠3, ∠6, ∠7

5. Which angles are obtuse angles?

∠1, ∠4, ∠5, ∠8

6. Find the measures of ∠1 and ∠2.

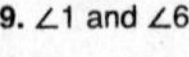

m∠1 = 140°; m∠2 = 40°

LESSON 8-3
Challenge
Transversal Hints

In the diagram below, $\overline{KL}$ is parallel to $\overline{MN}$, and $\angle JKL$ is congruent to $\angle JLK$.

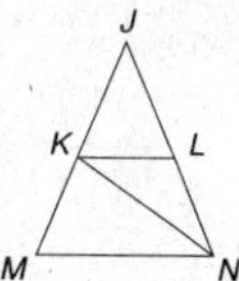

If $\angle JKL = 55°$, $\angle KNM = 33°$, and $\angle MJN = 70°$, find the following angle measures.

1. $\angle JMN$ — 55°
2. $\angle LKN$ — 33°
3. $\angle MKN$ — 92°
4. $\angle JLK$ — 55°
5. $\angle MNL$ — 55°
6. $\angle LNK$ — 22°

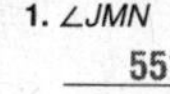

In the diagram below, line *A* is parallel to line *B*.

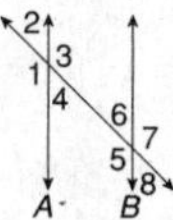

If $\angle 2 = x°$ and $\angle 7 = 3x°$, find the following angle measures.

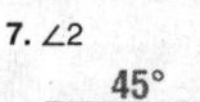

7. $\angle 2$ — 45°
8. $\angle 7$ — 135°
9. $\angle 3$ — 135°
10. $\angle 5$ — 135°
11. $\angle 8$ — 45°
12. $\angle 6$ — 45°

LESSON 8-3
Problem Solving
Angle Relationships

Write the correct answer.

In the drawing of the chair, the seat is parallel to the floor.

1. What is the measure of $\angle 1$?
   105°
2. What is the measure of $\angle 2$?
   75°
3. What is the measure of $\angle 3$?
   75°
4. What is the measure of $\angle 4$?
   85°

Choose the letter for the best answer.

The map shows the area around Falcon Park. Birch Street and Orchard Street are parallel to each other.

5. If $\angle 4$ measures 112°, what is the measure of $\angle 6$?
   A 112° (C) 68°
   B 22° D 108°
6. Which two angles are vertical angles?
   (F) $\angle 2$ and $\angle 3$ H $\angle 2$ and $\angle 4$
   G $\angle 2$ and $\angle 6$ J $\angle 2$ and $\angle 5$
7. If $\angle 10$ measures 87°, what is the measure of $\angle 9$?
   A 77° C 87°
   (B) 93° D 103°
8. Which is a transversal to Birch and Orchard streets?
   (F) Maple Street H Oak Street
   G Elm Street J Falcon Park
9. If $\angle 4$ measures 112°, what is the measure of $\angle 1$?
   A 22° C 108°
   B 68° (D) 112°

LESSON 8-3
Reading Strategies
Vocabulary Development

Lines that never meet and are the same distance apart are called **parallel lines.** The lines on sheet music are similar to parallel lines.

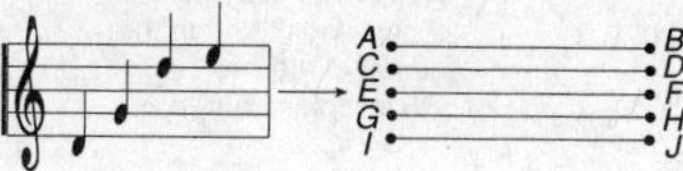

$\overline{EF} \parallel \overline{GH}$

Read: "Segment *EF* is parallel to segment *GH*."

1. Identify another pair of parallel line segments in the figure above.
   Possible answer: *AB* and *CD*
2. How would you describe parallel lines?
   Possible answer: Parallel lines are the same distance apart and never meet.

Lines that meet to form square corners are called **perpendicular lines.**

Angles formed by these square corners measure 90°. $\overrightarrow{WY} \perp \overleftrightarrow{ZW}$

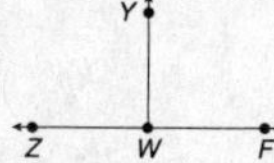

Read: "Ray *WY* is perpendicular to line *ZW*."

3. How can you tell if two lines, rays or segments are perpendicular?
   If the lines form square corners, they are 90° angles.
4. Identify another pair of perpendicular lines or rays in the figure.
   $\overrightarrow{WY} \perp \overrightarrow{WF}$

Lines that lie in different planes, but do not cross and are not parallel, are called **skew** lines.

$\overline{WC}$ and $\overline{TV}$ are skew.

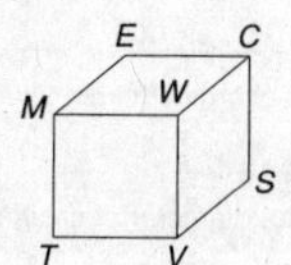

5. Identify two other line segments that are skew.
   Possible answer: $\overline{TV}$ and $\overline{ME}$ are skew.

LESSON 8-3
Puzzles, Twisters & Teasers
A Little Birdie Told Me!

Find the measure of each angle with a question mark. Each has a corresponding letter. Match the letters to the answers to solve the riddle. Assume lines that look parallel are parallel.

T 54°

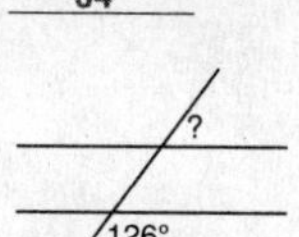

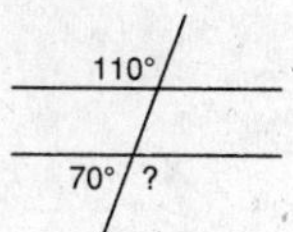

N 90°

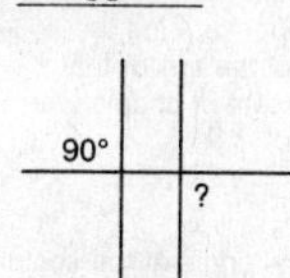

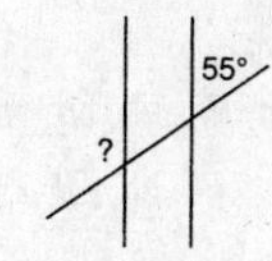

E 110°

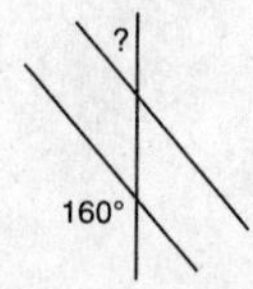

W 125°

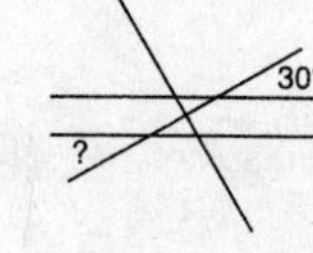

T 20°

?
160°

E 30°

30°
?

What do you give a sick bird?

A T(54) W(125) E(110) E T(20) M E(30) N(90) T

LESSON **8-4**

Practice A

Properties of Circles

Use circle C. Choose the letter for the best answer.

1. Which of the following is a radius?
   A $\overline{BD}$ Ⓒ $\overline{CE}$
   B $\overline{FB}$ D $\overline{GE}$

2. Which of the following is a diameter?
   F $\overline{FD}$ H $\overline{BC}$
   Ⓖ $\overline{AE}$ J $\overline{CG}$

3. Which of the following is a chord?
   Ⓐ $\overline{FD}$ C $\overline{BC}$
   B $\overline{CA}$ D $\overline{GF}$

Name the parts of circle K.

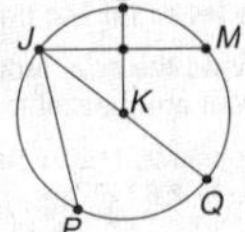

4. radii $\overline{KL}, \overline{KJ}, \overline{KQ}$
5. diameters $\overline{JQ}$
6. chords $\overline{JM}, \overline{JP}, \overline{JQ}$

Name the parts of circle T.

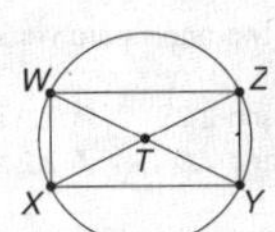
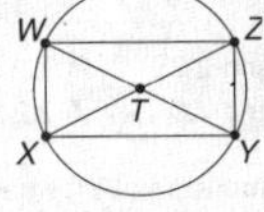

7. radii $\overline{WT}, \overline{XT}, \overline{YT}, \overline{ZT}$
8. diameters $\overline{WY}, \overline{XZ}$
9. chords $\overline{WX}, \overline{WY}, \overline{WZ}, \overline{XY}, \overline{XZ}, \overline{YZ}$

Use the circle graph.

10. The circle graph shows the uses of land in New Zealand. Find the central angle of the sector that shows the percent of land New Zealand uses for permanent pastures.

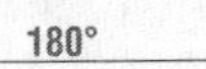
180°

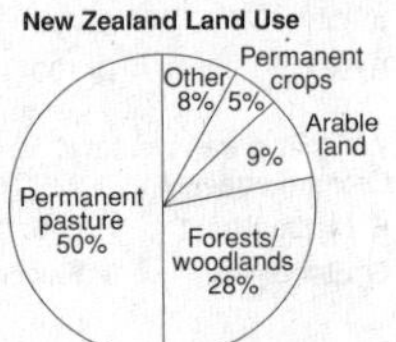
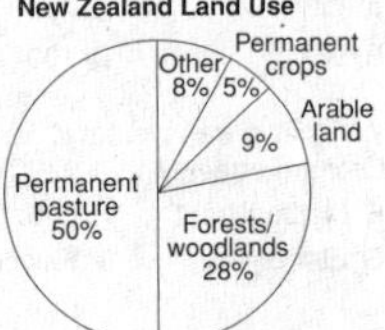

LESSON **8-4**

Practice B

Properties of Circles

Name the parts of circle A.

1. radii $\overline{AB}, \overline{AD}, \overline{AE}, \overline{AG}$
2. diameters $\overline{DE}$
3. chords $\overline{CF}, \overline{FG}, \overline{CB}, \overline{DE}$

Name the parts of circle H.

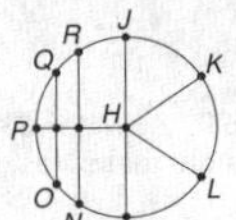

4. radii $\overline{HP}, \overline{HJ}, \overline{HK}, \overline{HL}, \overline{HM}$
5. diameters $\overline{JM}$
6. chords $\overline{OQ}, \overline{NR}, \overline{JM}$

Name the parts of circle C.

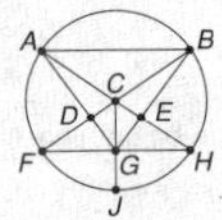

7. radii $\overline{CA}, \overline{CB}, \overline{CF}, \overline{CJ}, \overline{CH}$
8. diameters $\overline{AH}, \overline{FB}$
9. chords $\overline{AB}, \overline{FH}, \overline{AH}, \overline{FB}$

Name the parts of circle Z.

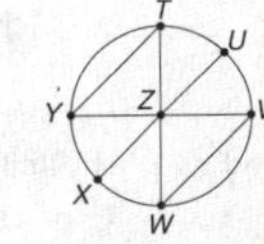
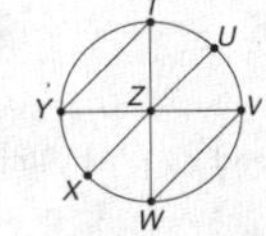

10. radii $\overline{ZY}, \overline{ZT}, \overline{ZU}, \overline{ZV}, \overline{ZW}, \overline{ZX}$
11. diameters $\overline{TW}, \overline{YV}, \overline{XU}$
12. chords $\overline{YT}, \overline{WV}, \overline{TW}, \overline{YV}, \overline{XU}$

Use the circle graph.

13. The circle graph shows the distribution of ethnic groups in New Zealand. Find the central angle measure of the sector that shows the percent of New Zealanders who are Maori.

34.92°

New Zealand Population
Maori 9.7%
Asian/other 7.4%
Other European 4.6%
Pacific Islander 3.8%
New Zealand European 74.5%

LESSON **8-4**

Practice C

Properties of Circles

Use the circle graph.

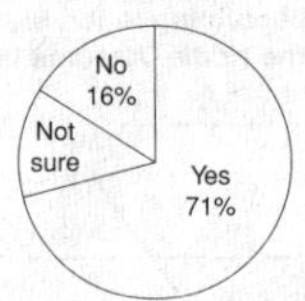
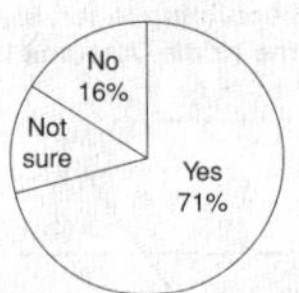

1. An Internet poll asked people if they believe that taking vitamins will make them live longer. The circle graph shows the response. Find the central angle of the sector that shows the percent of people who answered "No."

   57.6°

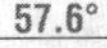

2. Find the central angle of the sector that shows the percent of people who answered "Yes."

   255.6°

3. Find the central angle of the sector that shows the percent of people who answered "Not sure."

   46.8°

Name the parts of circle N.

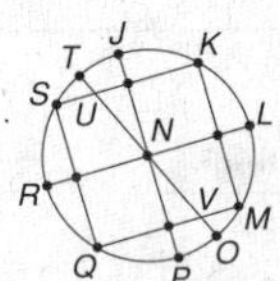

4. radii $\overline{NL}, \overline{NO}, \overline{NP}, \overline{NR}, \overline{NT}, \overline{NJ}$
5. diameters $\overline{RL}, \overline{JP}, \overline{TO}$
6. chords $\overline{SK}, \overline{KM}, \overline{MQ}, \overline{QS}, \overline{RL}, \overline{JP}, \overline{TO}$
7. If $\overline{SK} \parallel \overline{QM}$ and $m\angle NUK = 58°$, what is the measure of $\angle NVQ$?

   58°

8. If $m\angle NUK = 58°$, what is the measure of $\angle SUN$?

   122°

Use the clock for Exercises 9 and 10.

9. Find the central angle measure between the minute and hour hands on the clock.

   72°

10. Find the central angle measure between the second and hour hands on the clock.

   132°

LESSON **8-4**

Reteach

Properties of Circles

A chord is a line segment that connects 2 points on the circle. $\overline{JK}$ and $\overline{GH}$ are chords.

A diameter is a chord that passes through the center of the circle. $\overline{PR}$ is a diameter.

A circle is named by its center. This is circle O.

A radius is a line segment that connects a point on the circle with the center of the circle. $\overline{AO}$ and $\overline{BO}$ are radii.

Use circle A for Exercises 1–4.

1. $\overline{AB}$ is a radius.
2. $\overline{CD}$ is a diameter and chord.
3. $\overline{DE}$ is a chord.
4. $\overline{CA}$ is a radius.

Use circle Z for Exercises 5–7.

5. Name 3 radii of circle Z.

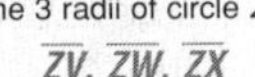
$\overline{ZV}, \overline{ZW}, \overline{ZX}$

6. Name 2 chords of circle Z.

   $\overline{YU}, \overline{VX}$

7. Name the diameter of circle Z.

   $\overline{VX}$

LESSON 8-4 **Challenge**
Arc Length

An *arc* is a portion of the circumference of a circle. On the circle below, arc *QR* corresponds to the part of the circumference whose endpoints are the radii forming central angle *QCR*.

A *minor arc* is less than 180° and is named by its endpoints. Arc *QR* is a minor arc. A *major arc* is greater than 180° and is named by its endpoints and one other point that lies on the arc. Arc *QSR* is a major arc.

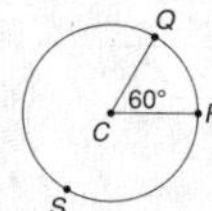

The degree measure of a minor arc is the measure of its central angle. The degree measure of a major arc is 360° minus the degree measure of the minor arc.

For circle *C* above, the measure of arc *QR* is 60°, and the measure of arc *QSR* is 360° − 60°, or 300°.

Find the degree measure of each arc.

1.

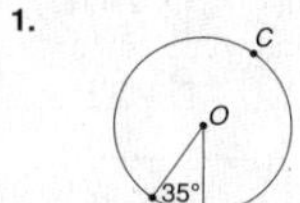
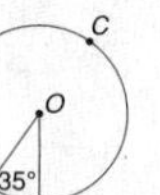

arc *BCA* 325°

2.

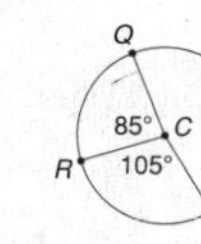

arc *QRS* 190°

3.

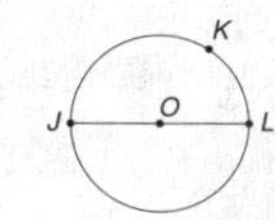

arc *JKL* 180°

4.

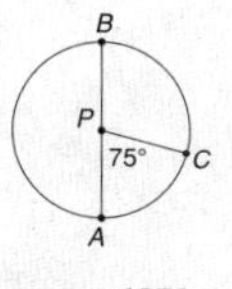

arc *BC* 105°

5.

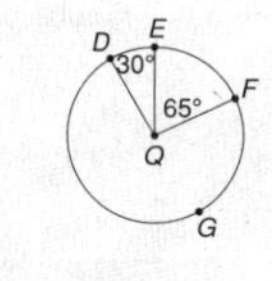

arc *DGF* 265°

6.

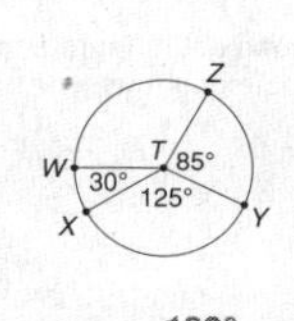

arc *WZ* 120°

LESSON 8-4 **Problem Solving**
Properties of Circles

Write the correct answer.

The circle graph shows the results of a survey in which people were asked to name their hobbies.

Americans Favorite Hobbies

Crafts 17%; Collections 3%; Sports 33%; 4% Computers; 9% Art/music; Other 22%; 12% Reading

1. If 1,000 people were surveyed, how many said that collecting things is their favorite hobby?

   30 people

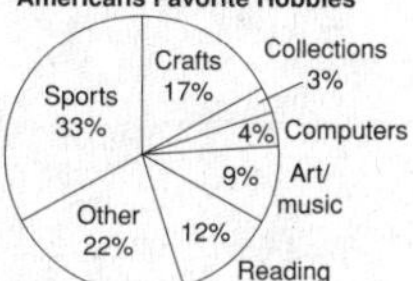

2. Find the central angle measure of the sector that shows the percent of people who named sports as their favorite hobby.

   118.8°

3. Find the central angle measure of the sector that shows the percent of people who named reading as their favorite hobby.

   43.2°

4. Find the central angle measure of the sector that shows the percent of people who like to do crafts as their favorite hobby.

   61.2°

Choose the letter for the best answer.

The circle graph shows the age breakdown of people who most enjoy snowboarding.

Snowboarding Popularity

18–24 22.4%; 12–17 39%; 7–11 19.4%; 55–up 1.4%; 35–54 6.5%; 25–34 11.3%

5. What is the central angle measure that shows the percent of 35–54-year-olds?
   A 5.1° C 23.4° (circled)
   B 6.5° D 40.7°

6. With which age group is snowboarding least popular?
   F 7–11 years H 25–34 years
   G 12–17 years J 55–up (circled)

7. With which age group is snowboarding most popular?
   A 7–11 years C 18–24 years
   B 12–17 years (circled) D 25–34 years

8. What is the central angle measure that shows the percent of 12–17-year-olds?
   F 39° H 108°
   G 140.4° (circled) J 180°

LESSON 8-4 **Reading Strategies**
Focus On Vocabulary

A **circle** is a closed figure. All points on the circle are the same distance from the **center point.** The center point is called the center of the circle.

The **diameter** is a line segment that passes through the center of a circle. It has its endpoints on the circle.

The **radius** is a line segment from the center of a circle to a point on the circle.

The **chord** is a line segment with endpoints on the circle.

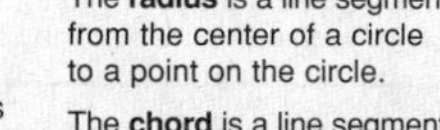
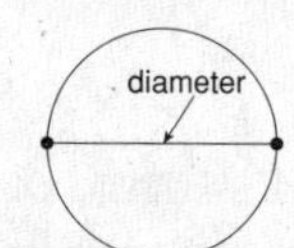

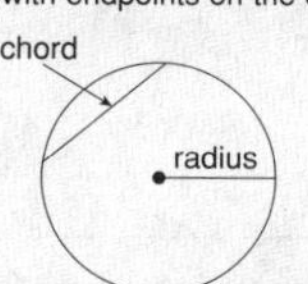

Answer each question.

1. What name is given to the line segment that passes through the center of a circle and has endpoints on the circle?

   diameter

2. What is the center point of a circle called?

   the center of the circle

3. What name is given to a line segment with endpoints on the circle?

   chord

4. Is a circle an open figure or a closed figure?

   closed

5. What is the name of the line segment that extends from the center of the circle to a point on the circle?

   radius

LESSON 8-4 **Puzzles, Twisters & Teasers**
Going in Circles

Across

1. In a plane, the set of all points that are the same distance from a given point
4. The part of a circle enclosed by two radii and an arc
5. The plural form of radius
7. Segment of a circle
8. Line segment that joins two points on a circle

Down

2. The point from which all points on a circle are equidistant
3. Angle formed by two radii
6. The straight line which passes through the center of a circle and has endpoints on the circle

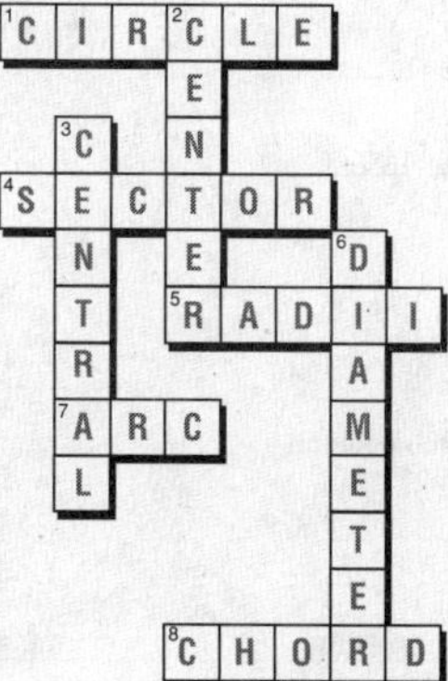

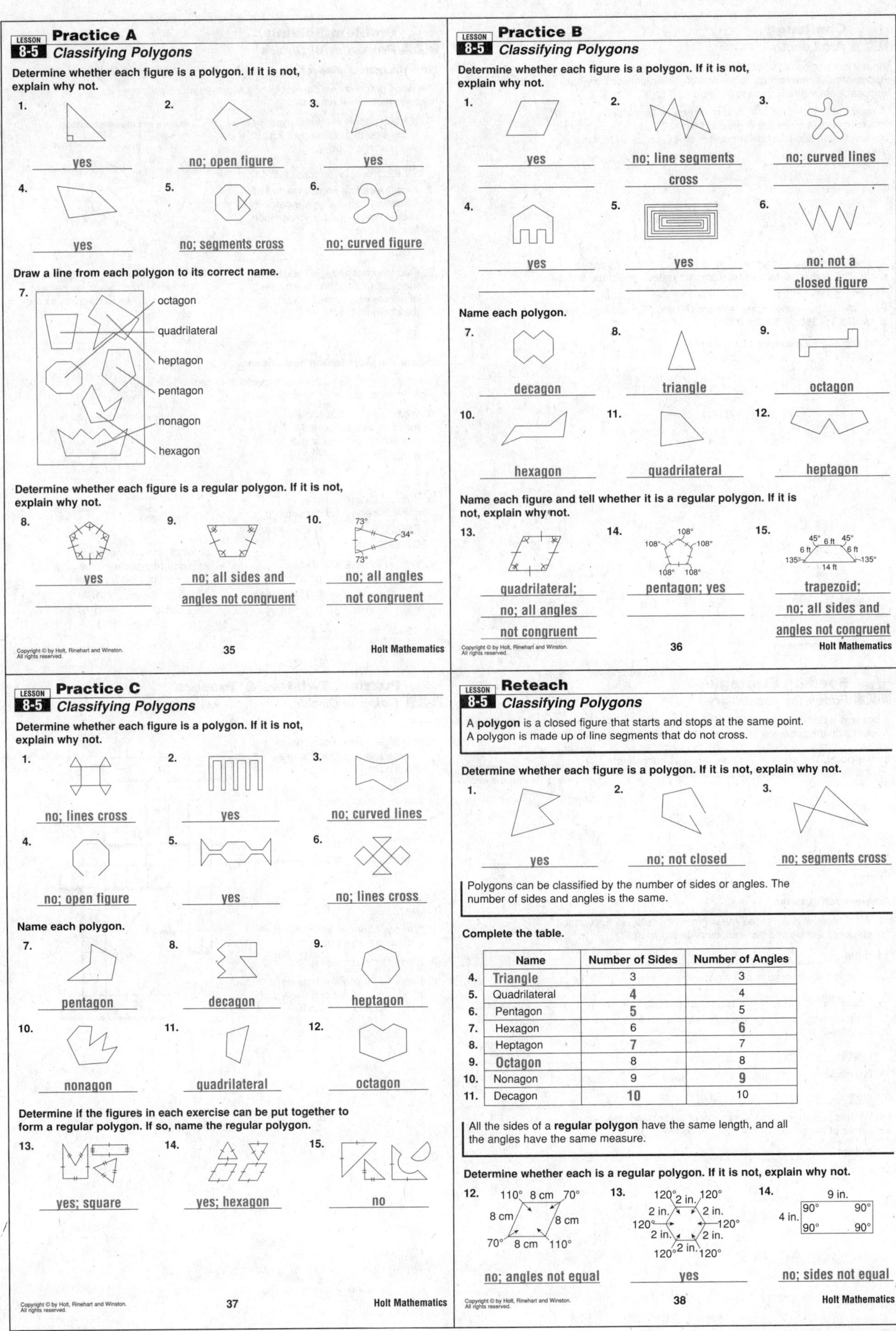

LESSON 8-5 Practice A
Classifying Polygons

Determine whether each figure is a polygon. If it is not, explain why not.

1. yes
2. no; open figure
3. yes
4. yes
5. no; segments cross
6. no; curved figure

Draw a line from each polygon to its correct name.

7. octagon, quadrilateral, heptagon, pentagon, nonagon, hexagon

Determine whether each figure is a regular polygon. If it is not, explain why not.

8. yes
9. no; all sides and angles not congruent
10. no; all angles not congruent

LESSON 8-5 Practice B
Classifying Polygons

Determine whether each figure is a polygon. If it is not, explain why not.

1. yes
2. no; line segments cross
3. no; curved lines
4. yes
5. yes
6. no; not a closed figure

Name each polygon.

7. decagon
8. triangle
9. octagon
10. hexagon
11. quadrilateral
12. heptagon

Name each figure and tell whether it is a regular polygon. If it is not, explain why not.

13. quadrilateral; no; all angles not congruent
14. pentagon; yes
15. trapezoid; no; all sides and angles not congruent

LESSON 8-5 Practice C
Classifying Polygons

Determine whether each figure is a polygon. If it is not, explain why not.

1. no; lines cross
2. yes
3. no; curved lines
4. no; open figure
5. yes
6. no; lines cross

Name each polygon.

7. pentagon
8. decagon
9. heptagon
10. nonagon
11. quadrilateral
12. octagon

Determine if the figures in each exercise can be put together to form a regular polygon. If so, name the regular polygon.

13. yes; square
14. yes; hexagon
15. no

LESSON 8-5 Reteach
Classifying Polygons

A **polygon** is a closed figure that starts and stops at the same point. A polygon is made up of line segments that do not cross.

Determine whether each figure is a polygon. If it is not, explain why not.

1. yes
2. no; not closed
3. no; segments cross

Polygons can be classified by the number of sides or angles. The number of sides and angles is the same.

Complete the table.

	Name	Number of Sides	Number of Angles
4.	Triangle	3	3
5.	Quadrilateral	4	4
6.	Pentagon	5	5
7.	Hexagon	6	6
8.	Heptagon	7	7
9.	Octagon	8	8
10.	Nonagon	9	9
11.	Decagon	10	10

All the sides of a **regular polygon** have the same length, and all the angles have the same measure.

Determine whether each is a regular polygon. If it is not, explain why not.

12. no; angles not equal
13. yes
14. no; sides not equal

LESSON 8-5 Challenge
The Diagonal Count

Consecutive vertices connect the sides of a polygon. A *diagonal* of a polygon is a segment that connects two nonconsecutive vertices.

diagonals

The figure shows two diagonals of a pentagon.

Other diagonals can be drawn in the pentagon from other vertices.

For each polygon below, find the total number of diagonals from all the vertices. Do not count the same diagonal more than once. *Hint:* Draw each figure and then draw all possible diagonals. Use different colors to distinguish the diagonals as the number of sides of the polygons increases.

1. Triangle 0 diagonals
2. Quadrilateral 2 diagonals
3. Pentagon 5 diagonals
4. Hexagon 9 diagonals
5. Heptagon 14 diagonals
6. Octagon 20 diagonals

LESSON 8-5 Problem Solving
Classifying Polygons

Write the correct answer.

The drawing shows a crown designed by a child in an arts and crafts class. The crown is composed of 4 different figures.

1 2 3 4

1. Name the polygon in figure 1.
   decagon
2. Name the polygon in figure 4.
   hexagon
3. Name the polygon in figure 3.
   pentagon
4. Is the crown a regular polygon? Explain.
   No, sides and angles not congruent.
5. Name the polygon formed by figures 2, 3, and 4.
   pentagon

Choose the letter for the best answer.

The box shows some basic shapes from a word processing tool bar.

1 2 3 4 5 6 7 8 9 10 11 12

6. Which figure is *not* a quadrilateral?
   A Figure 2　C Figure 4
   B Figure 3　(D) Figure 5
7. Which figure is *not* a polygon?
   F Figure 3　H Figure 11
   (G) Figure 9　J Figure 12
8. Which figure is a pentagon?
   A Figure 4　C Figure 10
   B Figure 5　(D) Figure 12
9. Which figure is a regular polygon?
   F Figure 1　(H) Figure 7
   G Figure 2　J Figure 8
10. Which figure is an octagon?
   (A) Figure 6　C Figure 11
   B Figure 10　D Figure 12

LESSON 8-5 Reading Strategies
Compare and Contrast

A **polygon** is a closed, plane figure formed by three or more line segments.

A **regular polygon** has sides of equal length, and all angle measures are the same.

Compare the regular and irregular polygons.

Regular Polygon　Irregular Polygon

Answer each question.

1. How many sides do both shapes have? 5 sides
2. How many angles do both shapes have? 5 angles
3. How are the two polygons alike?
   Both polygons have 5 sides and 5 angles
4. How are the two polygons different?
   Regular polygons have sides and angles of equal length, but irregular polygons do not.

Compare this regular octagon and regular pentagon.

Regular Pentagon　Regular Octagon

5. Compare the number of sides and angles in a pentagon to the number of sides and angles in an octagon.
   A pentagon has 5 sides and angles and an octagon has 8 sides and angles.
6. How are the octagon and pentagon alike?
   Possible answer: They each have sides of equal length and angles of equal measure.
7. How are the octagon and pentagon different?
   The number of sides and angles are different for each polygon.

LESSON 8-5 Puzzles, Twisters & Teasers
Bean Me Up, Scotty!

Decide whether or not each figure is a polygon. Circle the letters next to your answers. Unscramble the letters to solve the riddle.

1. polygon (L)　not polygon P
2. polygon (A)　not polygon D
3. polygon (M)　not polygon C
4. polygon P　not polygon (E)
5. polygon T　not polygon (A)
6. polygon (N)　not polygon E

What type of beans do llamas eat?

L L A M A　B E A N S

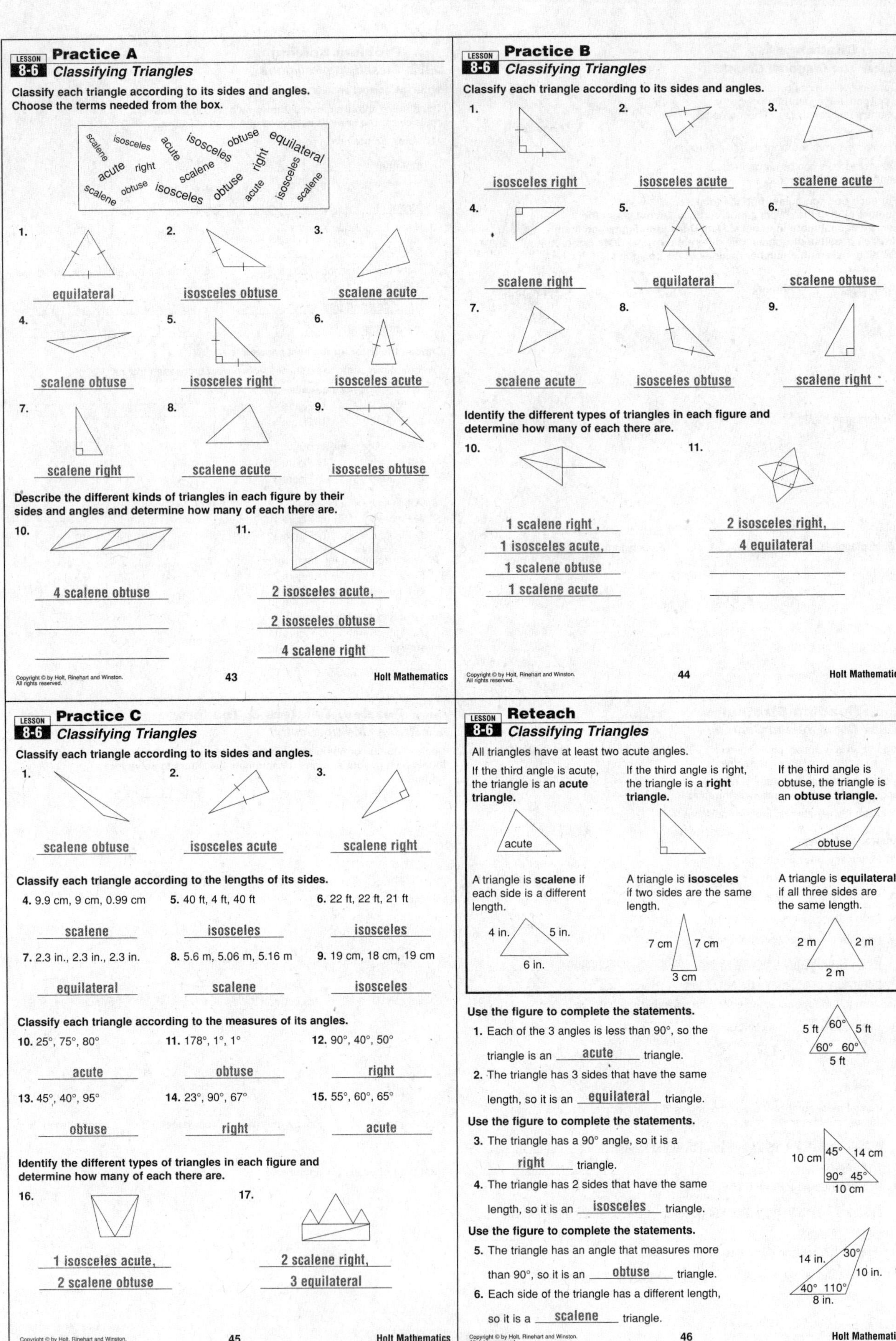

LESSON 8-6
Practice A
Classifying Triangles

Classify each triangle according to its sides and angles. Choose the terms needed from the box.

scalene isosceles acute isosceles obtuse equilateral acute right scalene right isosceles scalene obtuse isosceles obtuse acute scalene scalene

1. equilateral
2. isosceles obtuse
3. scalene acute
4. scalene obtuse
5. isosceles right
6. isosceles acute
7. scalene right
8. scalene acute
9. isosceles obtuse

Describe the different kinds of triangles in each figure by their sides and angles and determine how many of each there are.

10. 4 scalene obtuse
11. 2 isosceles acute, 2 isosceles obtuse, 4 scalene right

LESSON 8-6
Practice B
Classifying Triangles

Classify each triangle according to its sides and angles.

1. isosceles right
2. isosceles acute
3. scalene acute
4. scalene right
5. equilateral
6. scalene obtuse
7. scalene acute
8. isosceles obtuse
9. scalene right

Identify the different types of triangles in each figure and determine how many of each there are.

10. 1 scalene right, 1 isosceles acute, 1 scalene obtuse, 1 scalene acute
11. 2 isosceles right, 4 equilateral

LESSON 8-6
Practice C
Classifying Triangles

Classify each triangle according to its sides and angles.

1. scalene obtuse
2. isosceles acute
3. scalene right

Classify each triangle according to the lengths of its sides.

4. 9.9 cm, 9 cm, 0.99 cm scalene
5. 40 ft, 4 ft, 40 ft isosceles
6. 22 ft, 22 ft, 21 ft isosceles
7. 2.3 in., 2.3 in., 2.3 in. equilateral
8. 5.6 m, 5.06 m, 5.16 m scalene
9. 19 cm, 18 cm, 19 cm isosceles

Classify each triangle according to the measures of its angles.

10. 25°, 75°, 80° acute
11. 178°, 1°, 1° obtuse
12. 90°, 40°, 50° right
13. 45°, 40°, 95° obtuse
14. 23°, 90°, 67° right
15. 55°, 60°, 65° acute

Identify the different types of triangles in each figure and determine how many of each there are.

16. 1 isosceles acute, 2 scalene obtuse
17. 2 scalene right, 3 equilateral

LESSON 8-6
Reteach
Classifying Triangles

All triangles have at least two acute angles.

If the third angle is acute, the triangle is an **acute triangle.**	If the third angle is right, the triangle is a **right triangle.**	If the third angle is obtuse, the triangle is an **obtuse triangle.**
acute		obtuse
A triangle is **scalene** if each side is a different length.	A triangle is **isosceles** if two sides are the same length.	A triangle is **equilateral** if all three sides are the same length.
4 in. 5 in. 6 in.	7 cm 7 cm 3 cm	2 m 2 m 2 m

Use the figure to complete the statements.

5 ft 60° 5 ft 60° 60° 5 ft

1. Each of the 3 angles is less than 90°, so the triangle is an acute triangle.
2. The triangle has 3 sides that have the same length, so it is an equilateral triangle.

Use the figure to complete the statements.

10 cm 45° 14 cm 90° 45° 10 cm

3. The triangle has a 90° angle, so it is a right triangle.
4. The triangle has 2 sides that have the same length, so it is an isosceles triangle.

Use the figure to complete the statements.

14 in. 30° 10 in. 40° 110° 8 in.

5. The triangle has an angle that measures more than 90°, so it is an obtuse triangle.
6. Each side of the triangle has a different length, so it is a scalene triangle.

LESSON **8-6**
Challenge
Finding Shapes

1. How many triangles can you find?

11 triangles

2. How many triangles can you find?

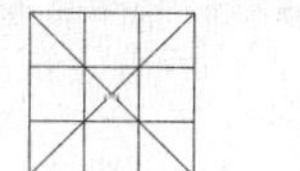

32 triangles

3. How many quadrilaterals can you find?

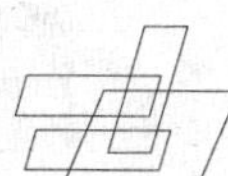

20 quadrilaterals

4. How many triangles and quadrilaterals can you find?

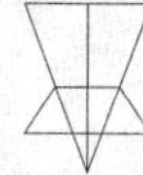

11 triangles and 15 quadrilaterals

Name a shape in the diagram as described. **Answers will vary. Possible answers are given.**

5. a right isosceles triangle

△*ABL*

6. a quadrilateral containing 2 right angles

quadrilateral *DEFO*

7. a right scalene triangle

△*CAL*

8. quadrilateral containing an obtuse angle and no right angles

quadrilateral *KMSR*

9. a quadrilateral with 2 pairs of equal sides

quadrilateral *KQHI*

10. an obtuse triangle

△*RQO*

11. an isosceles triangle that is not a right triangle

△*MNO*

LESSON **8-6**
Problem Solving
Classifying Triangles

Write the correct answer.

Brian made the following drawing of his kite.

1. How many triangles are in the figure?

8 triangles

2. How many acute triangles are in the figure?

2 acute triangles

3. How many right triangles are in the figure?

4 right triangles

4. How many scalene triangles are in the figure?

6 scalene triangles

5. How many isosceles triangles are in the figure?

2 isosceles triangles

Choose the letter for the best answer.

The figure shows an architect's design for a large window.

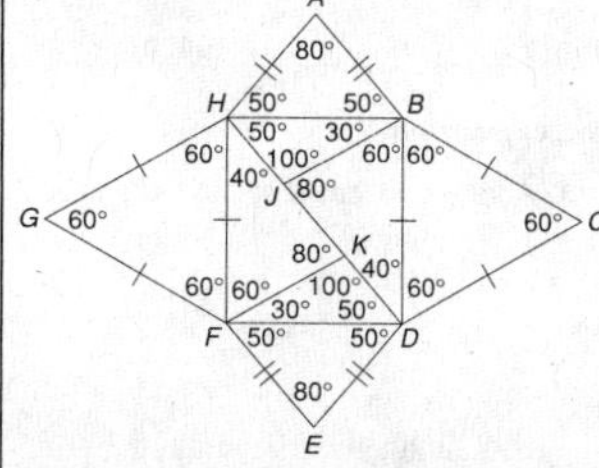

6. Which triangle is an equilateral triangle?
   A △*ABH*
   Ⓒ △*FGH*
   B △*BDH*
   D △*DHF*

7. Which triangle is a right triangle?
   F △*FHK*
   H △*DKF*
   G △*DEF*
   Ⓙ △*DBH*

8. Which triangle is an isosceles triangle?
   A △*DBJ*
   C △*HBJ*
   Ⓑ △*DEF*
   D △*DHF*

9. Which triangle is a scalene triangle?
   F △*ABH*
   H △*BCD*
   Ⓖ △*HBJ*
   J △*DEF*

10. Which triangle is an obtuse triangle?
    Ⓐ △*DKF*
    C △*JBD*
    B △*HKF*
    D △*ABH*

LESSON **8-6**
Reading Strategies
Make A Drawing

This chart will help you compare triangles by angles and by sides.

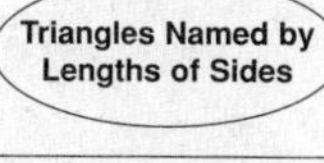

Triangles Named by Lengths of Sides	Triangles Named by Angle Measures
Equilateral Triangle All three sides are equal.	**Acute Triangle** All three angles measure less than 90°.
Isosceles Triangle Two sides are equal.	**Right Triangle** One angle measures 90°.
Scalene Triangle None of the sides are equal.	**Obtuse Triangle** One angle measures more than 90°.

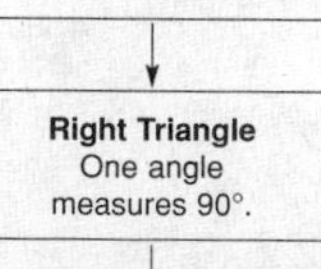

This drawing shows:
- One angle measures 90° → right triangle
- Two sides are equal. → isosceles triangle

The drawing shows an isosceles right triangle.

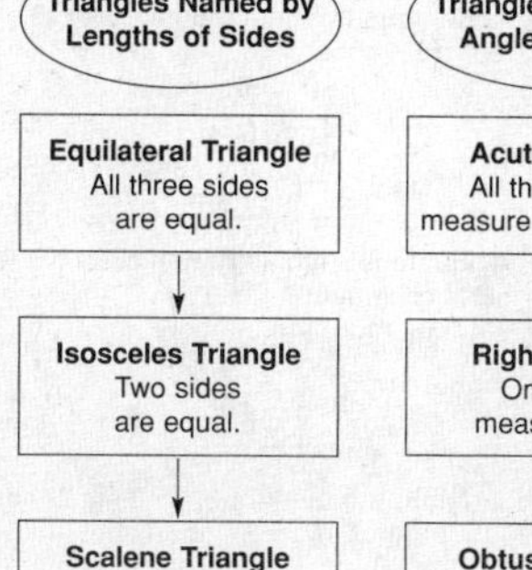

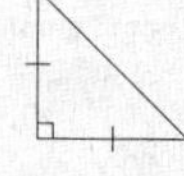

Draw the following triangles in the space provided.

1. A scalene obtuse triangle

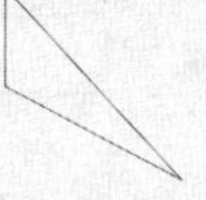

2. An equilateral triangle

LESSON **8-6**
Puzzles, Twisters & Teasers
You Are Correct, Sir!

Find and circle words from the word list in the word search (horizontally, vertically or diagonally). Find a word that answers the riddle. Circle it and write it on the line.

triangle	scalene	isosceles	equilateral	acute
obtuse	right	side	angle	measure

I N C O R R E C T L Y X
S O B T U S E X R C M E
O W E R T Y U L I E E R
S C A L E N E Q A M A T
C A C T G V U I N H S U
E C R I G H T O G F U P
L U W E R A N G L E R M
E T C V B S I D E B E Q
S E Q U I L A T E R A L

What is the only word in the dictionary that is spelled incorrectly?

Incorrectly

LESSON 8-7
Practice A
Classifying Quadrilaterals

1. List the five major special quadrilaterals.

parallelogram, rhombus, rectangle, square, trapezoid

Give all of the names that apply to each quadrilateral.

2. parallelogram, rhombus, rectangle square

3. trapezoid

4. parallelogram

Give the name that best describes each quadrilateral.

5. rhombus

6. rectangle

7. trapezoid

Draw each figure. If it is not possible to draw, explain why.

8. A rhombus that is also a square.

Possible answer is shown

9. A trapezoid that is also a rectangle.

It is impossible to draw. A trapezoid has exactly one pair of parallel sides, while a rectangle has two.

LESSON 8-7
Practice B
Classifying Quadrilaterals

Give all of the names that apply to each quadrilateral. Then give the name that best describes it.

1. parallelogram, rectangle; rectangle

2. parallelogram; parallelogram

3. parallelogram rectangle; rectangle

4. trapezoid; trapezoid

5. parallelogram, rhombus; rhombus

6. parallelogram, square, rectangle rhombus; square

Draw each figure. If it is not possible to draw, explain why.

7. A rectangle that is not a parallelogram.

It is impossible to draw. All rectangles are parallelograms. A rectangle has two pairs of parallel sides, as does a parallelogram.

8. A rectangle that is not a square.

LESSON 8-7
Practice C
Classifying Quadrilaterals

Give all of the names that apply to each quadrilateral. Then give the name that best describes it.

1. parallelogram; rectangle; rectangle

2. parallelogram, rhombus; rhombus

3. trapezoid

Draw each figure. If it is not possible to draw, explain why.

4. a rhombus that is not a rectangle

Possible answer is shown.

5. a trapezoid that is also a square

It is impossible to draw. A trapezoid has exactly one pair of parallel sides, while a square has two.

Name the types of quadrilaterals with each property.

6. exactly one pair of parallel sides trapezoid

7. four congruent sides rhombus, square

8. Graph the points $A(3, 3)$, $B(-1, 5)$, $C(-5, -1)$, and $D(1, -4)$ and draw quadrilateral $ABCD$. What quadrilateral did you draw?

trapezoid

9. Graph the points $W(1, 5)$, $X(3, -2)$, $Y(1, -5)$, and $Z(-1, 2)$ and draw quadrilateral $WXYZ$. What quadrilateral did you draw?

parallelogram

LESSON 8-7
Reteach
Classifying Quadrilaterals

Quadrilaterals are polygons that have 4 sides. In the figures below, arrows are used to indicate parallel sides, and tick marks are used to indicate congruent sides.

- A **trapezoid** is a quadrilateral with only one pair of parallel sides.
- A **rhombus** is a parallelogram with all four sides congruent.
- A **parallelogram** is a quadrilateral with two pairs of parallel sides.
- A **square** is a rectangle with all four sides congruent.
- A **rectangle** is a parallelogram with all four angles congruent.

Use the figures to complete the statements.

1. The figure is a trapezoid since it has one pair of parallel sides.

2. The figure is a parallelogram and a rhombus since it has two pair(s) of parallel sides, and four sides are congruent.

3. The figure is a parallelogram and a rhombus and a rectangle and a square since it has two pair(s) of parallel sides, four angles are congruent, and four sides are congruent.

LESSON 8-7

Challenge

It's Only a Quadrilateral!

Match each picture with the list showing which quadrilaterals are found in the picture.

1.
2.
3.

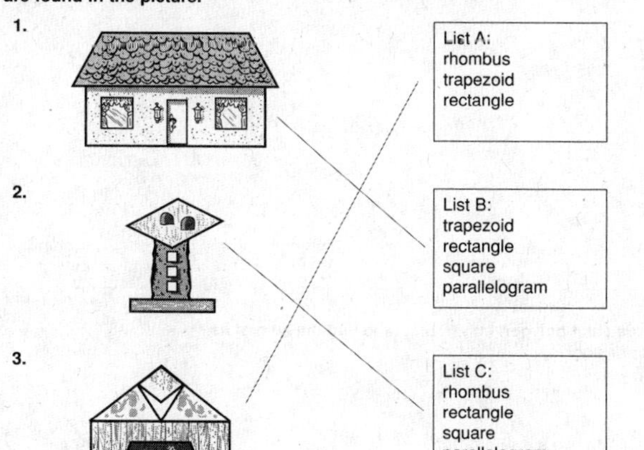

List A:
rhombus
trapezoid
rectangle

List B:
trapezoid
rectangle
square
parallelogram

List C:
rhombus
rectangle
square
parallelogram

4. Draw a design on the rug using only the following quadrilaterals: trapezoid, rectangle, and rhombus. You can use as many of each figure as you like, but all must be used at least once.

Answers will vary, but student's designs should include only trapezoids, rectangles, and rhombi, no parallelograms that are not rectangles.

LESSON 8-7

Problem Solving

Classifying Quadrilaterals

Write the correct answer.

1. A garden has 2 pairs of congruent sides. The congruent sides are not adjacent, and the angles are not all congruent. What is the shape of the garden?

   parallelogram

2. A poster advertising an art show has 1 pair of parallel sides. All the angles are different, and none of the sides are congruent. What is the shape of this poster?

   trapezoid

3. In a math report, Ming said that since a rhombus has 4 equal sides and a square has 4 equal sides, all rhombuses are squares. Is she correct? Explain.

   No, a rhombus does not always have 4 right angles.

4. A student was asked to give an example of a polygon that is not a quadrilateral and an example of a quadrilateral that is not a polygon. Is this possible? Why or why not?

   No, all quadrilaterals are polygons.

Choose the letter for the best answer.

This is a drawing of Piet Mondrian's painting *Composition 8.*

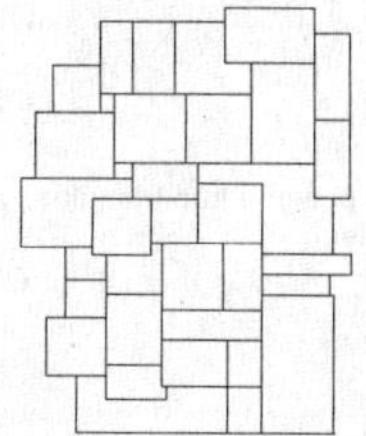

5. Which type of quadrilateral do you see most in the painting?
   A rhombus
   (C) rectangle
   B trapezoid
   D square

6. Which type of quadrilateral does not appear to be in the painting?
   F rectangle
   G parallelogram
   H square
   (J) trapezoid

7. Which statement is true about the painting?
   A There are no parallelograms.
   B There are only rectangles.
   C There are only quadrilaterals.
   (D) None of the above

8. Which statement is true about the painting?
   (F) There are no triangles.
   G There are only squares.
   H There are no squares.
   J None of the above

LESSON 8-7

Reading Strategies

Use A Diagram

Quadrilaterals are plane figures that have four sides and four angles. All the figures inside the largest region are quadrilaterals.

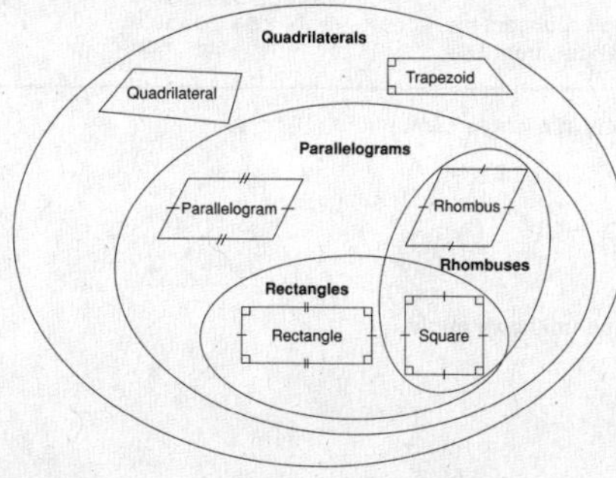

Parallelograms have two sets of opposite sides that are parallel and congruent. Opposite angles are also congruent. All the figures inside the next largest region are parallelograms.

Rectangles are quadrilaterals and parallelograms. They also have four right angles. **Rhombuses** are quadrilaterals and parallelograms with sides of equal length.

Use the diagram above to help you complete each question.

1. All parallelograms are quadrilaterals, true or false? **true**
2. Identify a quadrilateral that is not a parallelogram. **Possible answer: trapezoid**
3. Identify a quadrilateral that is also a parallelogram.. **Possible answers: rhombus, square, rectangle**
4. All parallelograms are rectangles, true or false? **false**
5. Identify a rectangle with four equal sides. **square**
6. A rhombus is a rectangle, true or false? **false**

LESSON 8-7

Puzzles, Twisters & Teasers

Charge!

Fill in the blanks to complete each statement. Use the letters above each number to solve the riddle below.

C O N G R U E N T means "same size and shape." (1)

A **P A R A L L E L O G R A M** has two pairs of parallel sides. (2)

A **T R A P E Z O I D** has only one pair of parallel sides. (3)

Q U A D R I L A T E R A L S are four-sided polygons that can have more than one name. (4)

A **T R I A N G L E** is a closed figure with three sides. (5)

A **R E C T A N G L E** has four right angles. (6)

A **C I R C L E** is a figure without angles or straight lines. (7)

A **S Q U A R E** has four congruent sides and four right angles. (8)

A **R H O M B U S** has four congruent sides and no right angles. (9)

Sides that are **A D J A C E N T** have a shared vertex. (10)

How do you stop a rhinoceros from charging?

Take away his **C R E D I T C A R D**
1 2 3 4 5 6 7 8 9 10

LESSON 8-8
Practice A
Angles in Polygons

Find the unknown angle measure in each polygon. Choose the letter for the best answer.

1. 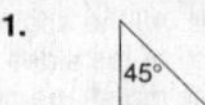

Ⓐ 45° C 90°
B 55° D 135°

2. 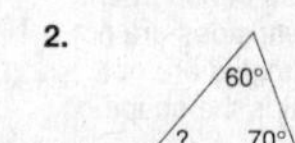

F 40° H 60°
Ⓖ 50° J 70°

3. 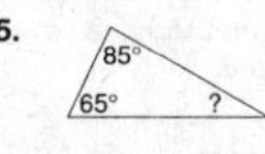

A 313° Ⓒ 113°
B 170° D 27°

4. 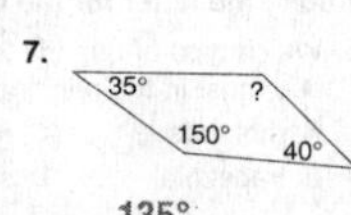

F 46° H 140°
Ⓖ 65° J 245°

Find the unknown angle measure in each polygon.

5.  30°

6. 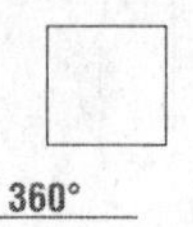 46°

7. 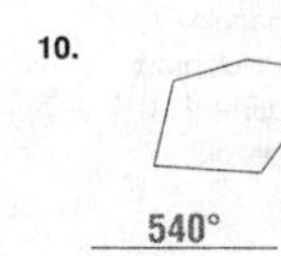 135°

Divide each polygon into triangles to find the sum of its interior angles.

8. 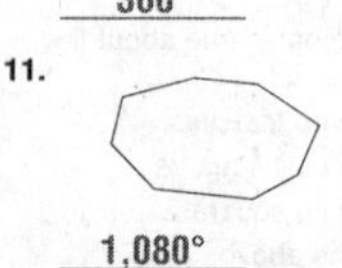 360°

9. 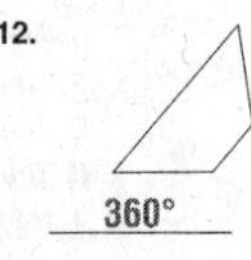 720°

10. 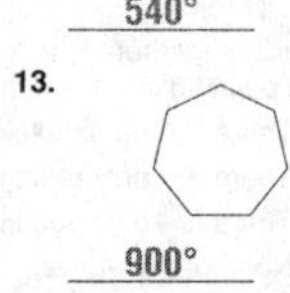 540°

11. 1,080°

12. 360°

13. 900°

LESSON 8-8
Practice B
Angles in Polygons

Find the unknown angle measure in each polygon.

1. 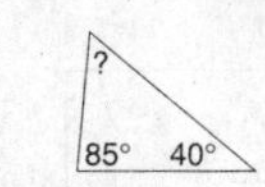 55°

2. 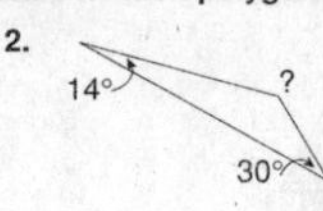 136°

3. 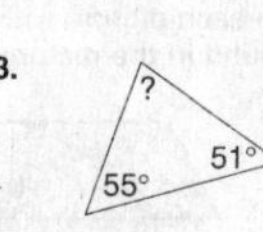 74°

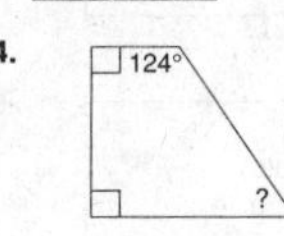

4. 56°

5. 115°

6. 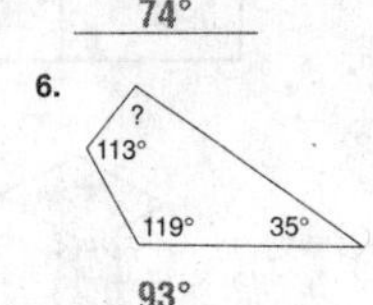 93°

Divide each polygon into triangles to find the sum of its interior angles.

7.  1,080°

8.  360°

9. 540°

10. 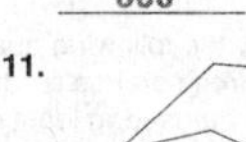 360°

11. 720°

12.  1,440°

13. A stop sign has the shape of a regular octagon. What is the sum of the interior angles of a stop sign?

1,080°

LESSON 8-8
Practice C
Angles in Polygons

Find the unknown angle measure in each polygon.

1. 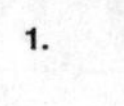 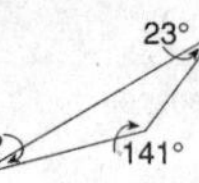 16°

2. 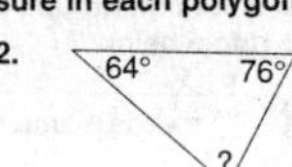 40°

3. 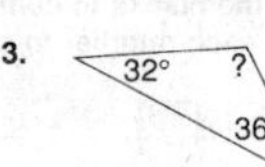 112°

4. 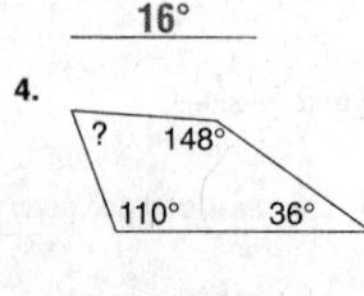 66°

5. 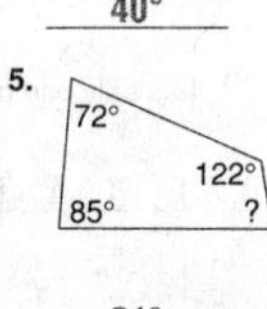 81°

6. 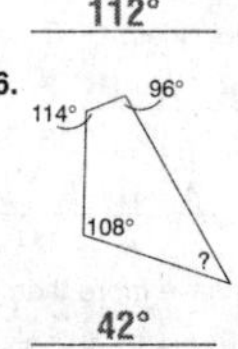 42°

Divide each polygon into triangles to find the sum of its interior angles.

7.  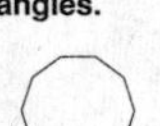 1,440°

8. 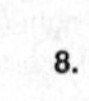 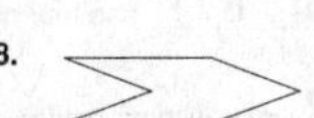 720°

9. 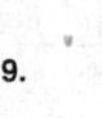 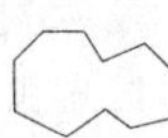 1,620°

Find the measure of the third angle in each triangle, given two angle measures. Then classify the triangle.

10. 99°, 34° — 47°; obtuse

11. 12°, 90° — 78°; right

12. 44°, 66° — 70°; acute

13. Mitsue has 3 planks that are each 4 feet long. If she uses the planks to form a triangular flower bed, what will be the measure of each angle in the triangle? 60°

LESSON 8-8
Reteach
Angles in Polygons

The sum of the measures of the three angles of any triangle is 180°.

To find the missing angle, subtract the sum of the two given angles from 180°.

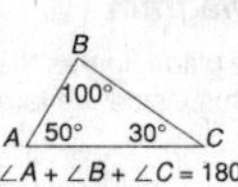

$\angle A + \angle B + \angle C = 180°$
$50° + 100° + 30° = 180°$

Find the measure of ∠N in triangle LMN.

1. $\angle N =$ 180° $- (74° + 57°)$
2. $\angle N =$ 180° $- 131°$
3. $\angle N =$ 49°

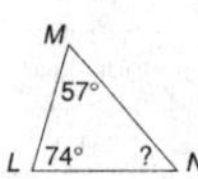

Find the measure of the unknown angle.

4. 103°

5. 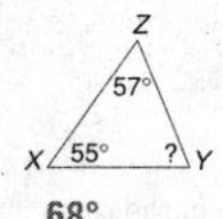 68°

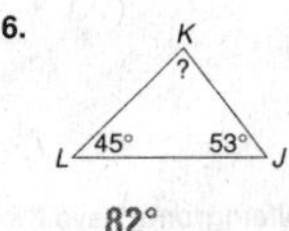

6. 82°

Use the figures to complete Exercises 7–10.

	Figure	Number of Sides	Number of Triangles
7.	Quadrilateral	4	2
8.	Pentagon	5	3
9.	Hexagon	6	4

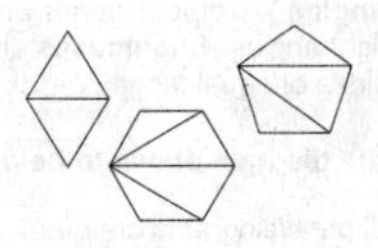

10. The number of triangles is always two less than the number of sides of the figure.

Find the sum of the interior angles.

11.  180° • 2 = 360°

12. 180° • 4 = 720°

13. 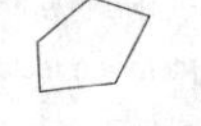 180° • 3 = 540°

LESSON 8-8 **Challenge**
Shape Up

Draw a polygon based on each description. Label the measures of all angles. Name the polygon. Answers may vary. Possible answers are given.

1. A figure has vertices *A, B, C, D,* and *E.* Sides *AB* and *BC* are 8 units in length. ∠*A* = ∠*C* = 130°. Side *CD* is parallel to *AE,* and each side is 10 units in length. ∠*D* = ∠*E* = 90°

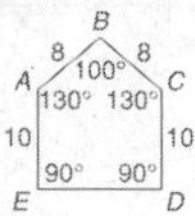

pentagon

2. A figure has vertices *W, X, Y,* and *Z.* Sides *WZ, WX,* and *XY* are each 12 units in length. Sides *WX* and *YZ* are parallel but are not the same length. ∠*W* = ∠*X* = 120°

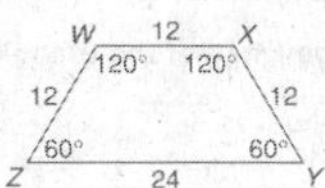

trapezoid

3. A figure has vertices *M, N, O, P, Q,* and *R.* Sides *MN* and *QP* are parallel, and each is 15 units in length. ∠*M* = ∠*N* = ∠*P* = ∠*Q* = 150°

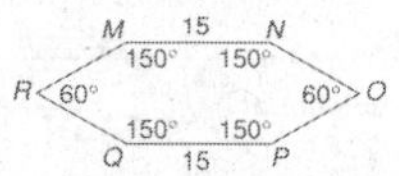

hexagon

4. A figure has vertices *A, B, C,* and *D.* *AB* = *BC* = *CD* = *AD.* Side *AB* is parallel to *CD,* and side *AD* is parallel to *BC.* ∠*B* and ∠*D* = 125°

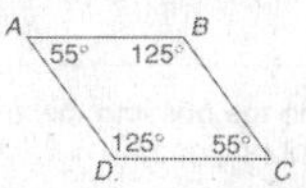

rhombus

5. A figure with vertices *E, F,* and *G* has perpendicular sides *EF* and *FG.* ∠*E* = ∠*G*

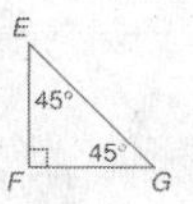

right triangle

6. A figure has vertices *Q, R, S, T.* Side *QR* is parallel to *ST,* and side *QT* is parallel to *RS.* ∠*Q* = 113°

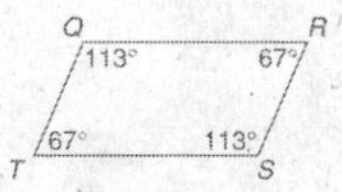

parallelogram

LESSON 8-8 **Problem Solving**
Angles in Polygons

Write the correct answer.

These are some common street signs.

1. The interior angles of which sign have a sum of 360°?
   one-way sign
2. What is the sum of the interior angles of a stop sign?
   1,080°
3. The interior angles of which sign have a sum of 540°?
   school crossing
4. If a yield sign is a regular polygon, what is the measure of each angle?
   60°

Choose the letter for the best answer.

5. A triangle with 3 unequal sides is a scalene triangle. Scalene triangles also have 3 unequal angles. Which angles can form a scalene triangle?
   A 50°, 60°, 65°
   (B) 50°, 60°, 70°
   C 110°, 40°, 40°
   D 20°, 20°, 40°
6. A design on a belt shows a row of irregular pentagons. Inside each figure, two of the angles each measure 60°. Another angle measures 270°. If the remaining angles are equal, what are their measures?
   F 60° each **(H)** 75° each
   G 120° each **J** 150° each
7. What is the missing angle measure of a triangle with angles 62° and 72°?
   A 52° **C** 54°
   B 34° **(D)** 46°
8. On a door is a sign that says "Welcome To Our Home." The sign is in the shape of a regular octagon. What is the measure of one of the interior angles of the sign?
   (F) 135° **H** 80°
   G 65° **J** 150°

LESSON 8-8 **Reading Strategies**
Make Predictions

Angles inside a polygon are called **interior angles.**

The sum of the interior angles of any triangle always equals 180°.

To find the sum of interior angles for other polygons:

- Divide the polygon into triangles.
- Draw all possible diagonals from one vertex.
- Multiply the number of triangles formed by 180°.

55°
35°
35° + 55° + 90° = 180°

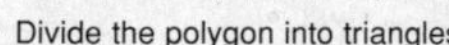

This quadrilateral can be divided into two triangles.

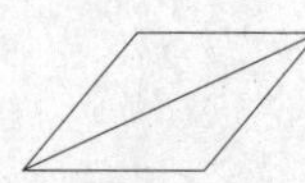

2 triangles formed

This pentagon can be divided into three triangles.

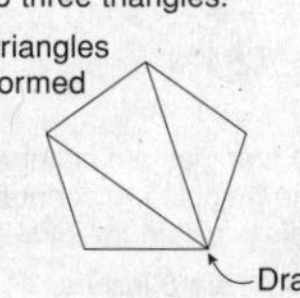

You can make a list of polygons and the number of triangles that can be made inside them.

Polygon	Number of Sides	Number of Triangles
Triangle	3	1
Quadrilateral	4	2
Pentagon	5	3
Hexagon	6	?
Septagon	7	?
Octagon	8	?

Use the pattern in the list to predict the number of triangles for the rest of the polygons.

1. How many triangles do you think can be formed inside a hexagon? 4
2. What helped you make your prediction?
   Possible answer: the pattern in the list
3. How many triangles do you predict can be formed inside a septagon? 5
4. How many triangles do you predict can be formed inside an octagon? 6

LESSON 8-8 **Puzzles, Twisters & Teasers**
What's Your Angle?

Find the unknown angle measure in each figure. Each answer has a corresponding letter. Use the letters to solve the riddle.

1. *t* = 37°

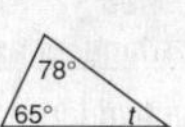

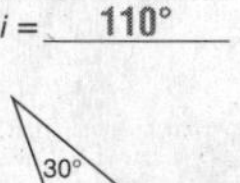

2. *n* = 60°

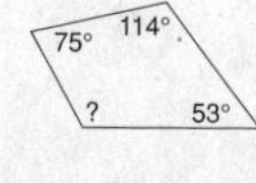

3. *i* = 110°

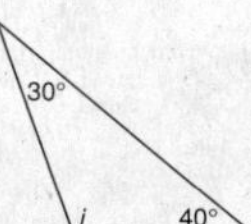

4. *a* = 118°

75°
114°
?
53°

5. *c* = 50°

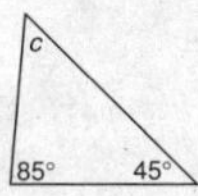

6. *u* = 30°

Why would you take a pencil to bed?

To draw the C(50) U(30) R T(37) A(118) I(110) N(60) S

LESSON 8-9
Practice A
Congruent Figures

Are there any congruent figures in each picture? If there are, describe them.

1.

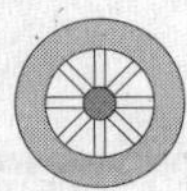

Yes; the spokes on the wheel are congruent.

2.

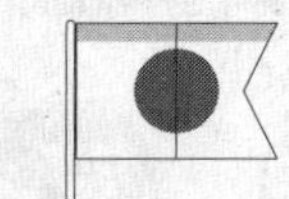

No

3.

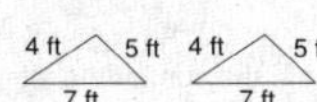

yes

4.

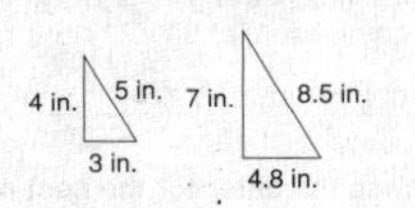

4 cm 12 cm 9 cm 4 cm 12 cm 9 cm

yes

5.

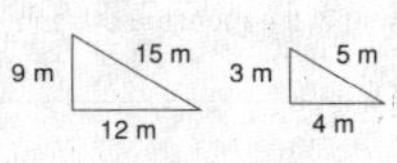

no

6.

4 in. 5 in. 3 in. 7 in. 8.5 in. 4.8 in.

no

Determine the missing measure in each set of congruent polygons.

7.

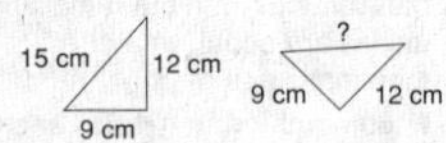

15 cm

8.

7 in. 82° 4 in. ? 60° 8 in. 4 in. 82° 7 in. 60° 38° 8 in.

38°

9.

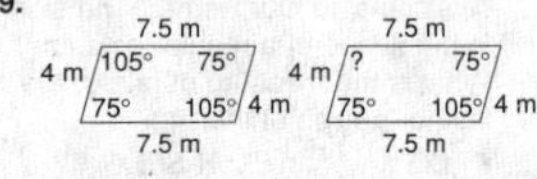

105°

10.

10 ft 13 ft 21 ft 18 ft 10 ft 13 ft ? 18 ft

21 ft

LESSON 8-9
Practice B
Congruent Figures

Identify any congruent figures.

1.

Yes; the short stripes are congruent and the stars are congruent.

2.

There are no congruent figures in the picture.

Determine whether the triangles are congruent.

3.

22 m 10 m 14 m 14 m 10 m 22 m

yes

4.

3 in. 3 in. 5 in. 7 in. 4 in. 4 in.

no

5.

14 cm 13 cm 16.5 cm 12 cm 18 cm 15 cm

no

6.

62 ft 34 ft 52 ft 34 ft 52 ft 62 ft

yes

Determine the missing measure or measures in each set of congruent polygons.

7.

21 in. 34 in. 52 in. ? 34 in. 21 in.

52 in.

8.

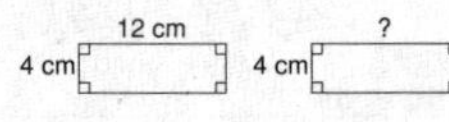

12 cm

9.

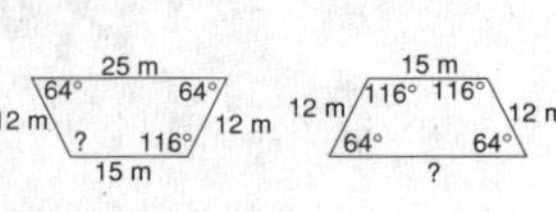

116°; 25 m

10.

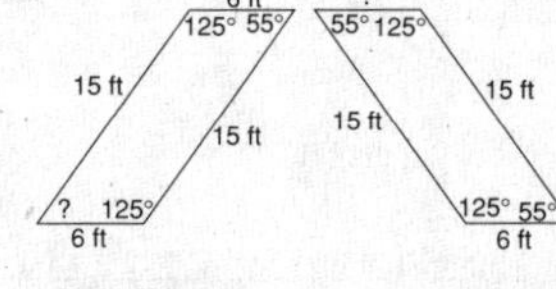

55°; 6 ft

LESSON 8-9
Practice C
Congruent Figures

Identify any congruent figures.

1.

There are no congruent figures in the picture.

2.

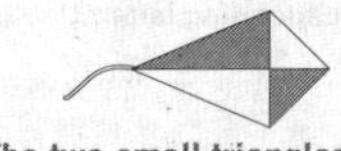

The two small triangles are congruent, and the two larger triangles are congruent.

Determine whether the triangles are congruent.

3.

22 m 24.6 m 11 m 11 m 24.6 m 22 m

yes

4.

6 in. 7 in. 4 in. 9 in. 8 in. 7 in.

no

Determine the missing measures in each set of congruent polygons.

5.

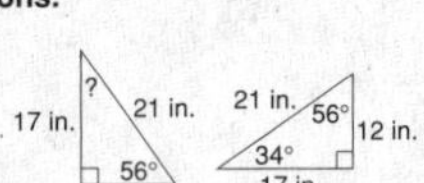

34°; 12 in.

6.

11 cm 11 cm ? 144° 84° 25 cm 25 cm 48° 48° 25 cm 25 cm 84° 144° 84° 11 cm ?

84°; 11 cm

7.

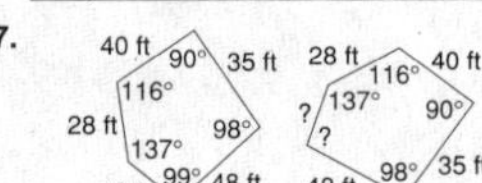

99°; 16 ft

8.

58 ft 105° 75° 52 ft 52 ft 78 ft 52 ft ? ? 58 ft 75° 78 ft

52 ft; 105°

9. Describe how to determine if two hexagons are congruent.

Two hexagons are congruent if corresponding sides and angles are equivalent.

LESSON 8-9
Reteach
Congruent Figures

When two polygons are congruent, the angles and the sides of one polygon are equivalent to the corresponding angles and sides of the other polygon.

Find the corresponding side or angle. *ABCD* ≅ *EFGH*

1. $\overline{AB} \cong$ $\overline{EF}$

2. $\angle DAB \cong$ $\angle HEF$

3. $\overline{GH} \cong$ $\overline{CD}$

4. $\angle GFE \cong$ $\angle CBA$

A 9 cm B 6 cm 8 cm D 14 cm C F 9 cm E 6 cm 8 cm H G 14 cm

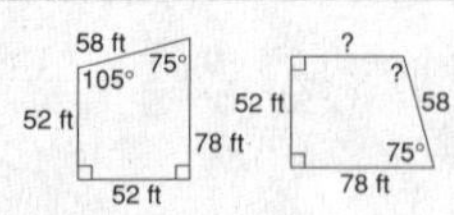

To show that two triangles are congruent, you can show that the three sides of one triangle are congruent to the three sides of the other triangle. This is called the *Side-Side-Side* rule.

$\overline{PQ} \cong \overline{ST}$ since both are 8 inches.

$\overline{QR} \cong \overline{TU}$ since both are 6 inches.

$\overline{PR} \cong \overline{SU}$ since both are 10 inches.

So, $\triangle PQR \cong \triangle STU$.

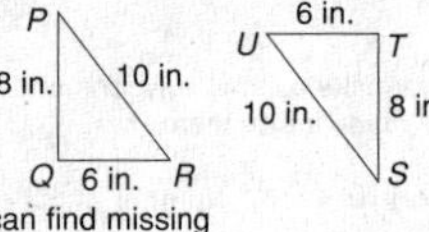

If you know that two figures are congruent, you can find missing measures in the figures.

- The corresponding angles are in the same position.
- The corresponding sides are in the same position.

Complete.

5. *JKLMN* ≅ *VWXYZ*

$\angle K \cong \angle W$, so the measure of $\angle W$ is 100°.

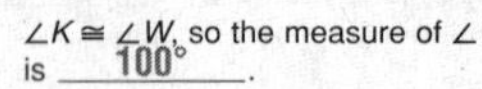

K 9 in. 100° 7 in. J 115° 110° L 6 in. 100° 115° ? N 8 in. M W 7 in. ? ? V X 110° 115° 7 in. 6 in. 115° 100° Y 8 in. Z

6. *JKLMN* ≅ *VWXYZ*

$\overline{LM} \cong \overline{XY}$, so the length of $\overline{LM}$ is 7 in.

7. What is the length of $\overline{VW}$? 9 in.

LESSON 8-9 Challenge
Congruent Presidents?

Find the figures in each row that appear to be congruent. Write their letters on the line next to the figures. Unscramble the letters to name the president.

1. 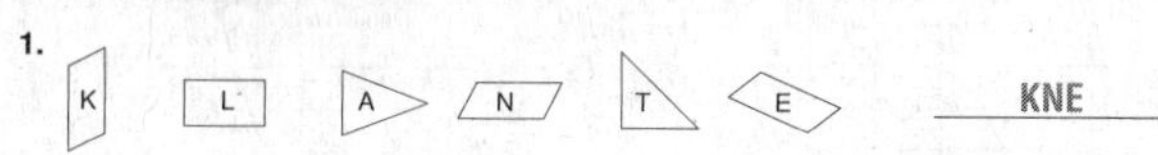

KNE

2. 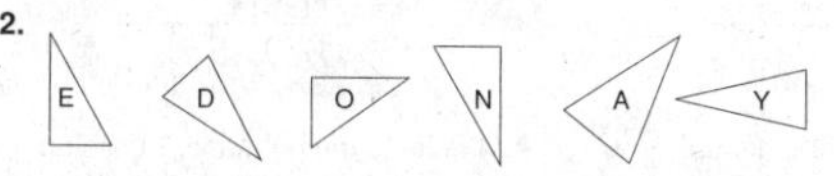

EN, DY

3. This president had a pony named Macaroni. Kennedy

4. 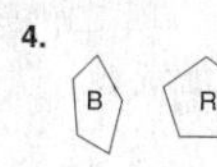 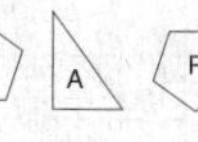 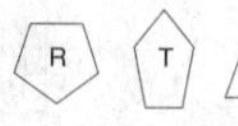 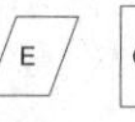 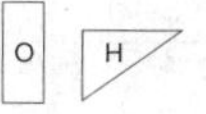 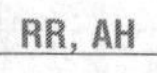

RR, AH

5. 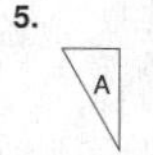 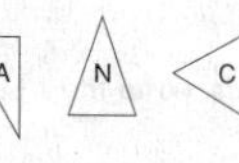 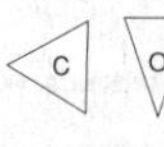  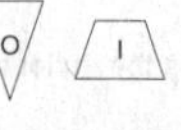 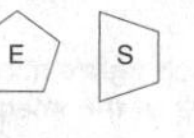 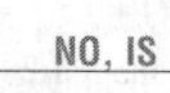

NO, IS

6. This president had a goat named Old Whiskers who used to pull the president's grandchildren in a cart. Harrison

7. 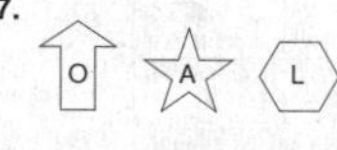 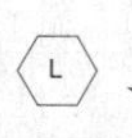 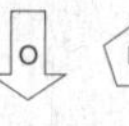 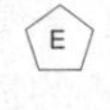 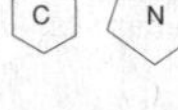 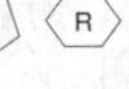

OO, CL

8. 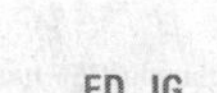

ED, IG

9. This president walked his raccoon named Rebecca on a leash at the White House. Coolidge

LESSON 8-9 Problem Solving
Congruent Figures

Write the correct answer.

The table shows the dimensions of regulation NBA and NCAA basketball courts, which are rectangular in shape.

Basketball Court and Lane Sizes

	NBA	NCAA
Court	94 ft by 50 ft	94 ft by 50 ft
Lane	16 ft by 19 ft	12 ft by 19 ft

1. Is an NBA court congruent to an NCAA court? Why or why not?

Yes, they are both rectangles and both 94 ft by 50 ft.

2. If the lane on an NBA court is the same shape as the lane on an NCAA court, are they congruent? Explain.

No, they are not the same size.

3. The lane on a WNBA court is 12 feet by 19 feet. If the lane is the same shape as that of the NBA lane, are they congruent? Why or why not?

No, they are not the same size.

Choose the letter for the best answer.

4. The parallelograms below are congruent. What is the measure of $\angle U$?

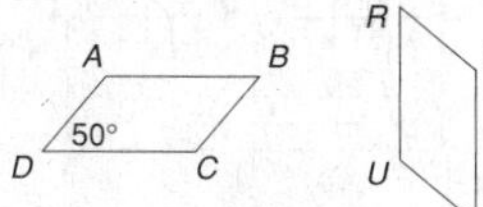

A 50°
(B) 130°
C 100°
D 260°

5. The Mexican flag is a rectangle divided into three vertical stripes of identical measures. Which of the following statements is true about the Mexican flag?

(F) It has 3 congruent rectangles.
G It has 3 rectangles that are not congruent.
H It has 4 congruent rectangles.
J It has 4 rectangles and none are congruent.

6. In $\triangle ABC$, $m\angle A = m\angle B$, and $m\angle C = 100°$. What are the measures of the angles in $\triangle DEF$ if it is congruent to $\triangle ABC$?

A 60°, 60°, 100°
B 20°, 60°, 100°
C 30°, 60°, 100°
(D) 40°, 40°, 100°

7. $\triangle JKL$ is congruent to $\triangle RST$. $\angle L$ and $\angle T$ are right angles. Which statement about the two triangles is *not* true?

F $m\angle K = m\angle S$
G $\overline{JK} \cong \overline{RS}$
(H) $m\angle J = m\angle L$
J $m\angle R + m\angle S = 90°$

LESSON 8-9 Reading Strategies
Graphic Organizer

This chart helps you understand congruence.

Definition

Two figures that have exactly the same size and shape are congruent.

Facts

- Corresponding sides are congruent.
- Corresponding angles are congruent.

Congruence

Examples **Non-examples**

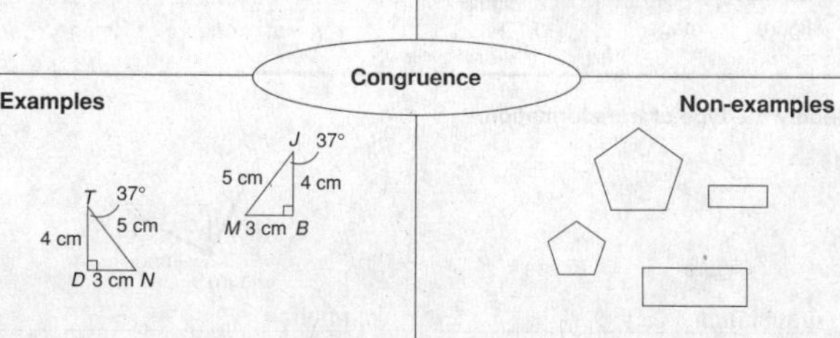

Use the chart to answer each question.

1. How can you tell if two figures are congruent?

They have the same size and same shape.

2. Are the rectangles congruent? Why or why not?

No; They are not the same size.

3. Compare the two triangles. Which side corresponds to TD?

JB

4. How long are $\overline{TD}$ and $\overline{JB}$?

4 centimeters

5. Which angle corresponds to angle *D*?

angle *B*

6. Are the two triangles congruent? Explain why or why not.

Yes; Possible answer: corresponding sides and corresponding angles are congruent.

LESSON 8-9 Puzzles, Twisters & Teasers
Hit the Nail on the Head!

Decide whether or not each pair of figures is congruent. Circle the letter above your answer. Unscramble the letters to solve the riddle.

1. 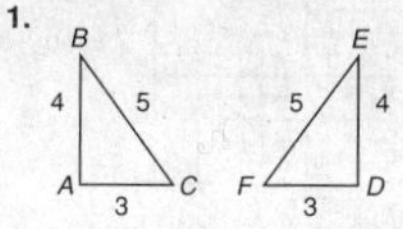

(T) congruent K not congruent

2. 62° $5\frac{2}{3}$ $6\frac{2}{3}$ 81° 5 127° 2 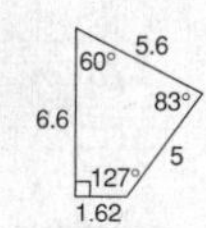

P congruent (H) not congruent

3. 120 m, 70 m, 70 m, 120 m — 120 m, 70 m, 70 m, 120 m

(M) congruent R not congruent

4. N, 6 in., 8 in., M, 7 in., O — 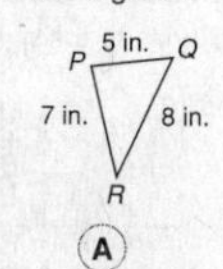

L congruent (A) not congruent

5. 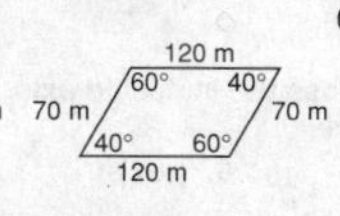

120 m 60° 40° 70 m 70 m 40° 60° 120 m

U congruent (B) not congruent

6. 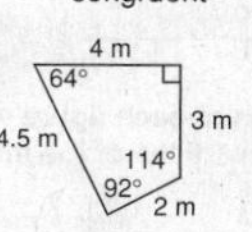

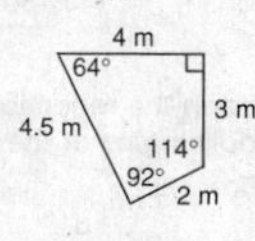

(I) congruent W not congruent

Which nail does a carpenter hate to hit?

His T H U M B N A I L

LESSON 8-10
Practice A
Translations, Reflections, and Rotations

Identify the transformation. Choose the letter of the best answer.

1.

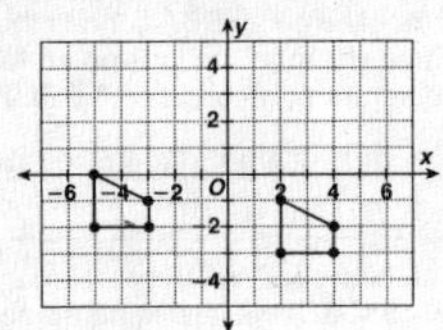

(A) translation
B reflection
C rotation

2.

A translation
(B) reflection
C rotation

Graph each translation.

3. 2 units to the left and 4 units down

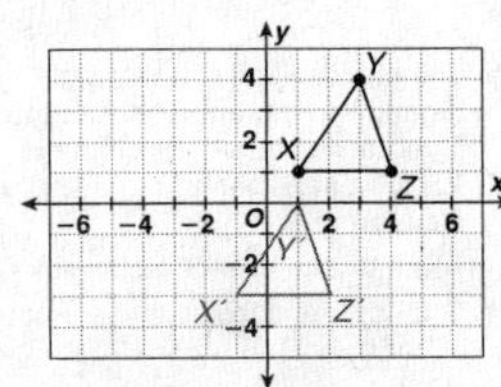

4. 3 units to the right and 2 units up

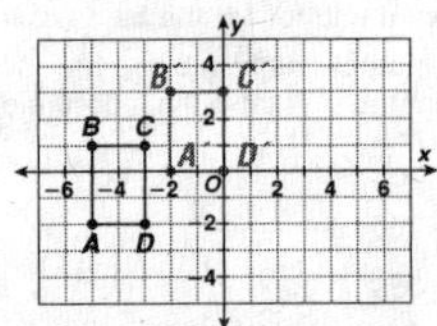

Follow the directions to graph each transformation.

5. Reflect $\triangle EFG$ across the x-axis.

6. Rotate $\triangle XYZ$ 180° about the vertex X.

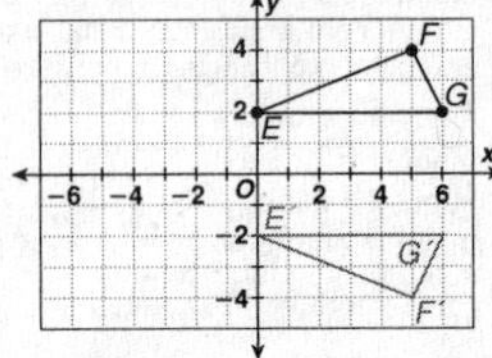

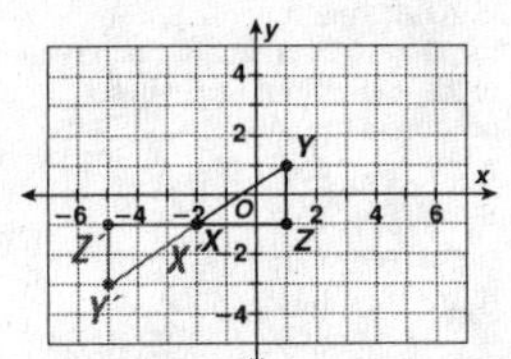

LESSON 8-10
Practice B
Translations, Reflections, and Rotations

Identify each type of transformation.

1.

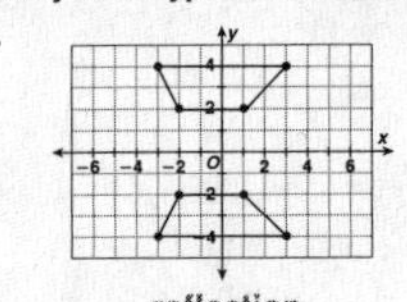

reflection

2.

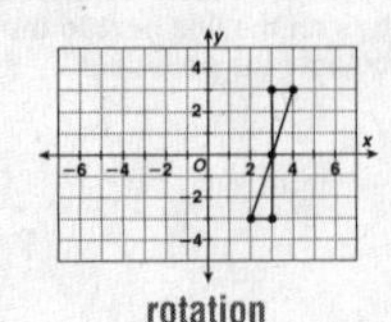

rotation

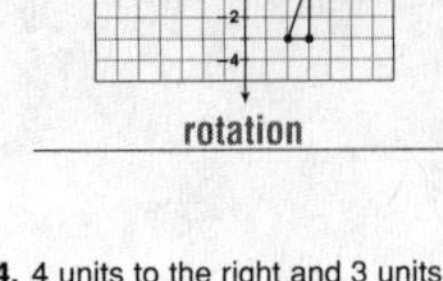

Graph each translation.

3. 5 units to the left and 2 units up

4. 4 units to the right and 3 units up

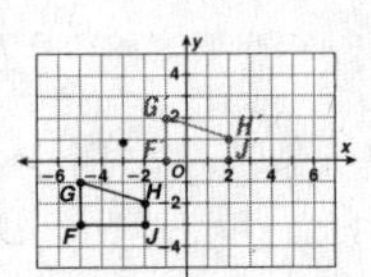

Graph the reflection of each figure across the indicated axis. Write the coordinates of the vertices of the image.

5. x-axis

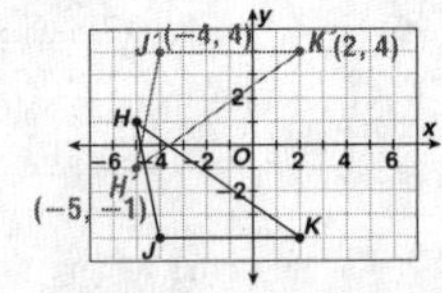

6. y-axis

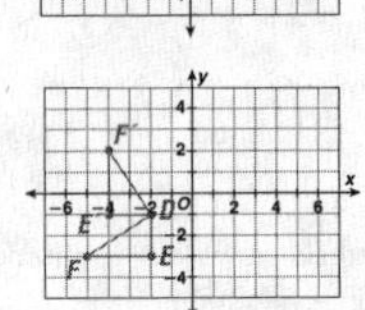

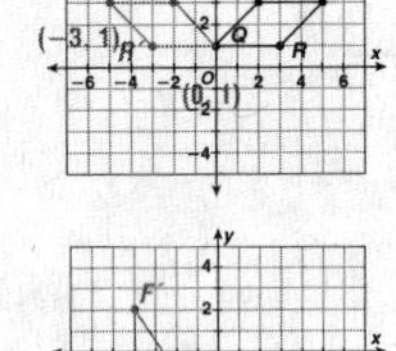

7. Triangle DEF has vertices at $D(-2, -1)$, $E(-2, -3)$, and $F(-5, -3)$. Rotate $\triangle DEF$ 90° clockwise about the vertex D.

LESSON 8-10
Practice C
Translations, Reflections, and Rotations

Identify each type of transformation.

1.

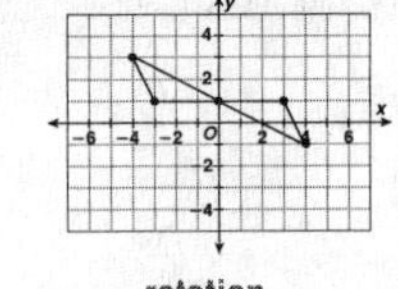

rotation

2.

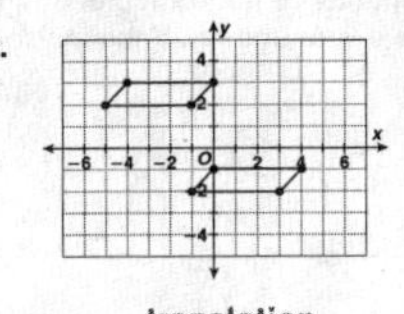

translation

Graph each translation.

3. 3 units to the right and 3 units down

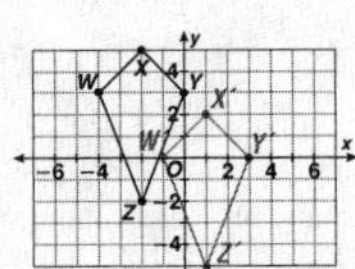

4. 4 units to the left and 5 units up

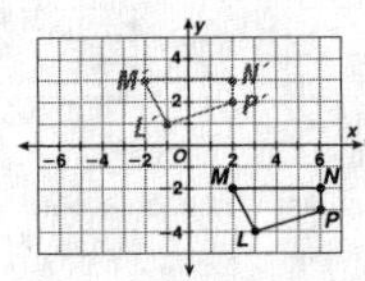

Graph the reflection of each figure across the indicated axis. Write the coordinates of the vertices of the image.

5. x-axis

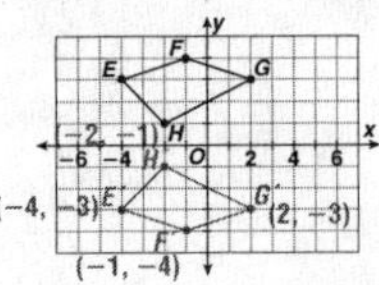

6. y-axis

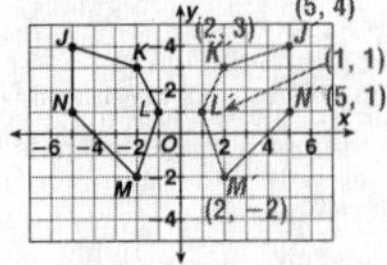

7. Triangle KLM has vertices at $K(1, -1)$, $L(4, -1)$, and $M(5, 1)$. Rotate $\triangle KLM$ 270° counterclockwise about the vertex K.

LESSON 8-10
Reteach
Translations, Reflections, and Rotations

A **translation** is a *slide* to a new position.

A **rotation** is a *turn* of the figure.

A **reflection** is a *flip* of the figure.

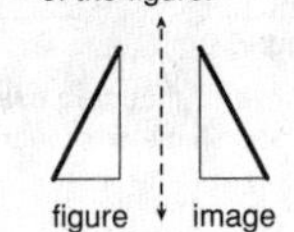

Identify the type of transformation.

1.

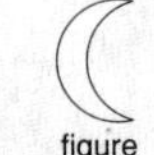

translation

2.

rotation

3.

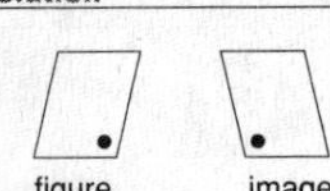

reflection

4.

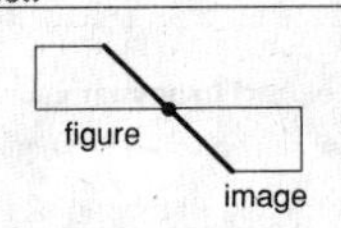

rotation

- You can translate figures in the coordinate plane.

$\triangle ABC$ is translated 3 units right and 3 units down.

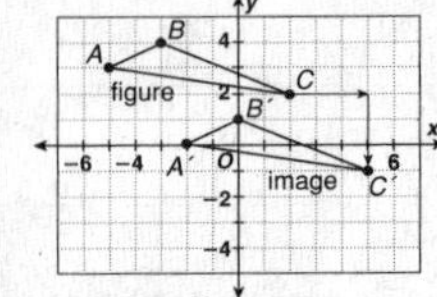

$\triangle DEF$ is translated 2 units left and 3 units up.

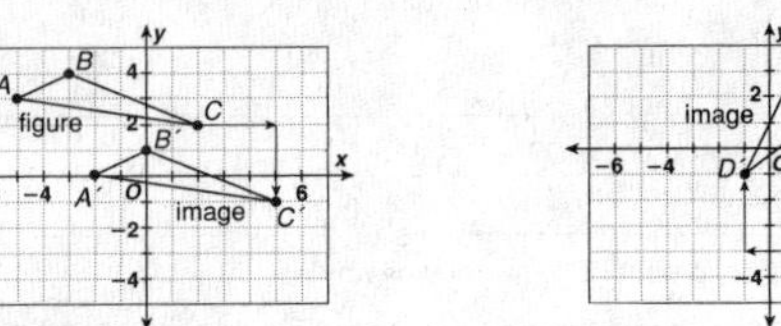

LESSON 8-10 **Reteach**

Translations, Reflections, and Rotations (continued)

5. When you translate a figure to the right, does the x-coordinate of the image increase or decrease? increase

6. When you translate a figure downward, does the y-coordinate of the image increase or decrease? decrease

- You can rotate a figure about a point in the coordinate plane.

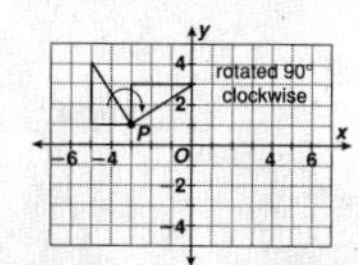

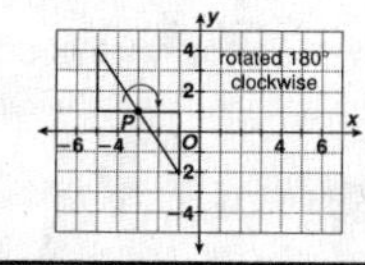

7. When you rotate a figure about a point, does the point of rotation move? no

- You can reflect a figure across a line.

△JKL is reflected across the x-axis. △PQR is reflected across the y-axis.

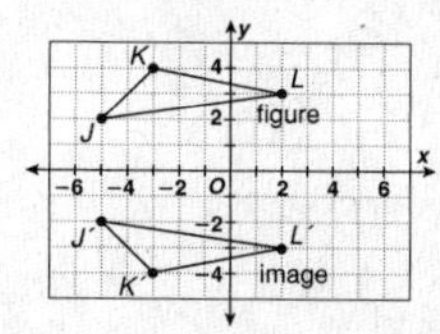

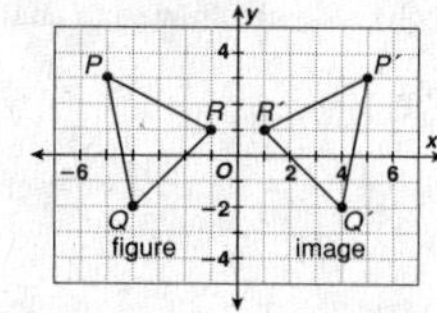

8. When you reflect a figure across the x-axis, do the x-coordinates change or remain the same? remain the same

9. When you reflect a figure across the x-axis, do the y-coordinates change or remain the same? change

10. When you reflect a figure across the y-axis, do the x-coordinates change or remain the same? change

11. When you reflect a figure across the y-axis, do the y-coordinates change or remain the same? remain the same

LESSON 8-10 **Challenge**

Follow That Transformation

You alone know the location of the secret treasure. You want to draw a map to direct your friend to the treasure, but you must do so in code. Use transformations to make the code.

First, draw a triangle on the coordinate plane below.

Next, list in order 5 transformations that must be done to the original figure. The treasure is located in the image of the last transformed triangle.

Then exchange your map with another student. Follow the directions and draw a triangle to show the location of the treasure. Possible answer given.

1. reflect across x-axis

2. rotate 180° about point (2, −3)

3. translate 4 units left and 10 units up

4. reflect across y-axis

5. translate 7 units left and 6 units down

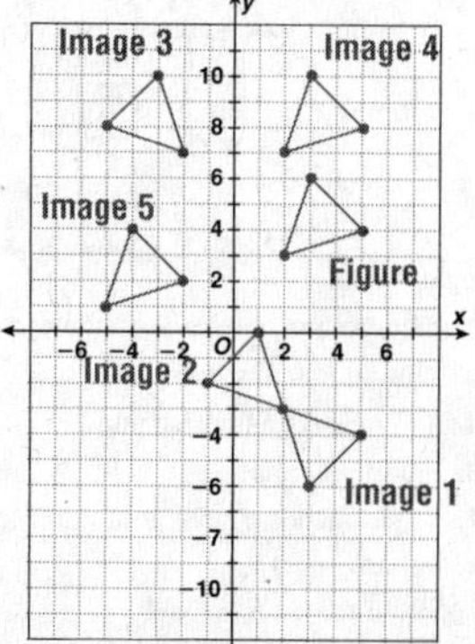

LESSON 8-10 **Problem Solving**

Translations, Reflections, and Rotations

Write the correct answer.

Clock 1

Clock 2

1. If you reflect the hands of clock 1 across a line from 12 to 6, what time will it show? 11:00

2. If you rotate the hour hand on clock 2 by 90° clockwise, what time will it be? 6:00

3. The hands on clock 1 show 7:00 after a transformation of one hand. What was the transformation? 180° rotation of the hour hand

4. The hands on clock 2 show 9:00 after a transformation. Name 2 different transformations that could produce this change. reflection across a line from 12 to 6, or rotating the hour hand 180°

Choose the letter for the best answer.

5. What transformation of triangle 1 created triangle 2?
   A translation 3 units right and 1 unit down
   (B) translation 8 units right and 1 unit down
   C rotation of 180° about the origin
   D reflection across the y-axis

6. If you rotate triangle 2 90° clockwise about vertex D, what will be the coordinates of the new triangle?
   (F) D′(3, 1), E′(7, 1), F′(3, −3)
   G D′(3, 1), E′(3, −3), F′(7, 1)
   H D′(3, 1), E′(−4, 1), F′(−3, 3)
   J D′(3, 1), E′(−3, 3), F′(−7, 1)

7. If you reflect triangle 1 across the x-axis, what will be the coordinates of the new triangle?
   A A′(5, 2), B′(5, 6), C′(1, 2)
   B A′(−5, 0), B′(-5, −4), C′(−1, 0)
   C A′(5, −2), B′(5, −6), C′(1, −2)
   (D) A′(−5, −2), B′(−5, −6), C′(−1, −2)

LESSON 8-10 **Reading Strategies**

Use Graphics

A **transformation** moves a figure but does not change its shape. The figure that moves is called the **image** of the original figure.

Here are three kinds of transformations.

- **Translation**
  Slide a figure to a new position.

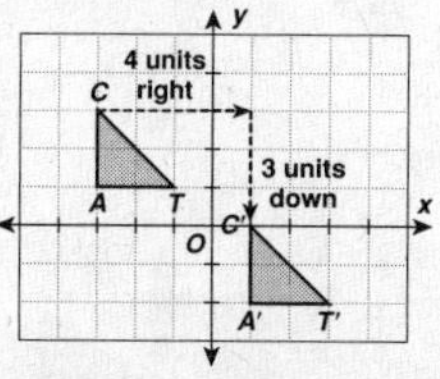

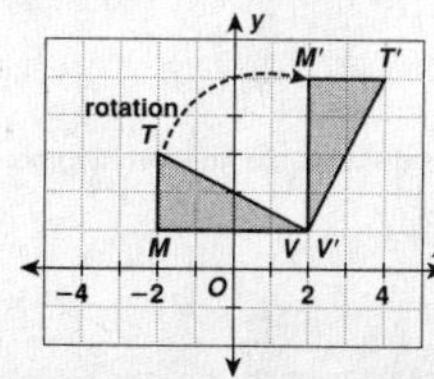

△CAT has moved four units to the right and three units down.

- **Reflection**
  Flip a figure across the y-axis.

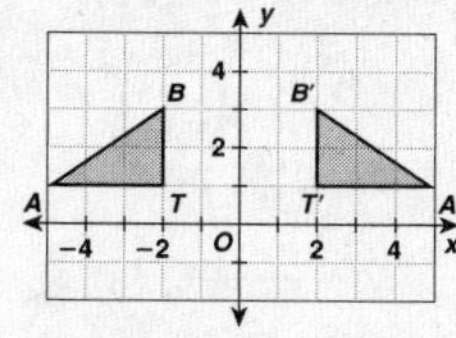

The triangle has been reflected across the y-axis.

- **Rotation**
  Turn a figure around a point.

△MTV has been rotated 90° clockwise around vertex V.

1. Translate △DEF six units to the right and two units down.

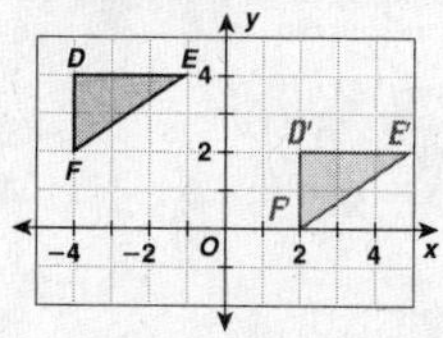

2. Reflect the figure across the y-axis.

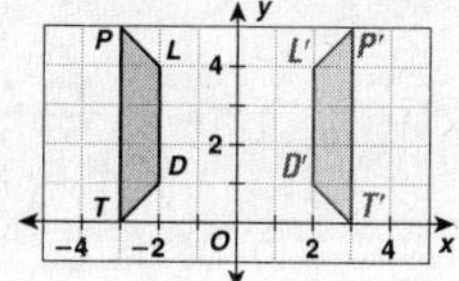

LESSON 8-10 Puzzles, Twisters & Teasers
Transformer!

Across

1. A way to produce a new image from a given one
6. Not moving
8. A reflection is a _____ image
9. _____ of reflection

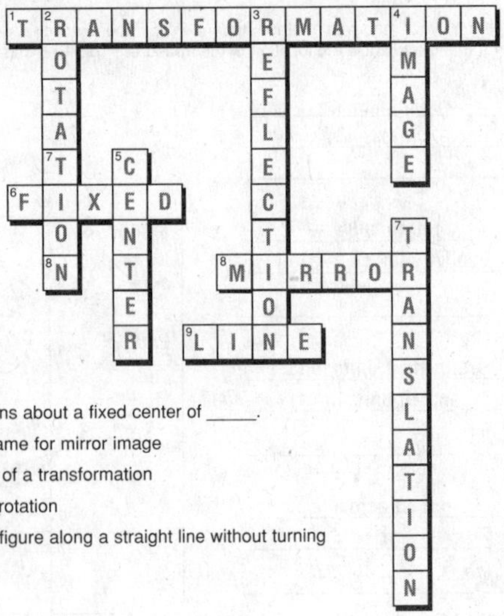

1 TRANSFORMATION; 2 ROTATION; 3 REFLECTION; 4 IMAGE; 5 CENTER; 6 FIXED; 7 TRANSLATION; 8 MIRROR; 9 LINE

Down

2. A figure turns about a fixed center of _____.
3. Another name for mirror image
4. The result of a transformation
5. _____ of rotation
7. To slide a figure along a straight line without turning

LESSON 8-11 Practice A
Symmetry

Draw all the lines of symmetry in each figure.

1.  A
2. 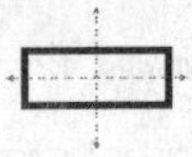
3.

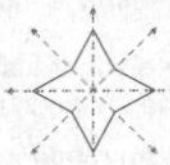

Tell how many lines of symmetry each flag has.

4. Scotland 2
5. Thailand 1
6. Chile 0

Using the given figure, draw each position.

7. Z Original; A 90° turn; Another 90° turn; Another 90° turn

8. How many times does the figure above show rotational symmetry within one full rotation?

2 times

How many times will each figure show rotational symmetry within one full rotation?

9.  4
10.  2
11.  4

LESSON 8-11 Practice B
Symmetry

Decide whether each figure has line symmetry. If it does, draw all the lines of symmetry.

1. 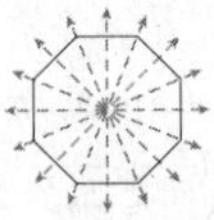
2. 
3.

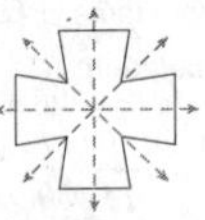

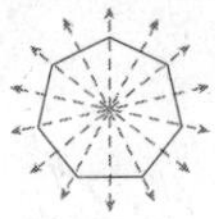

Find and draw all the lines of symmetry in each flag.

4. Iceland 
5. Nauru 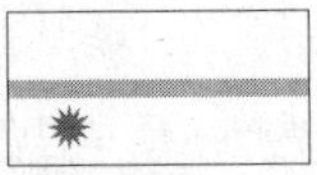 none
6. Burundi

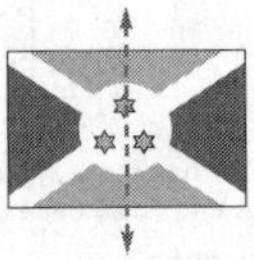

Tell how many times each figure will show rotational symmetry within one full rotation.

7.  2
8.  5
9.  3
10.  4
11.  5
12.  6

LESSON 8-11 Practice C
Symmetry

Decide whether each figure has line symmetry. If it does, draw all the lines of symmetry.

1. 
2.  8
3.

Find and draw all the lines of symmetry in each flag.

4. Great Britain 
5. Guyana 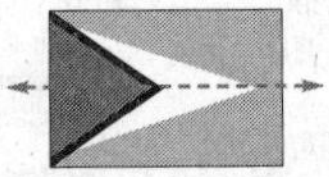
6. Trinidad and Tobago  none

Tell how many times each figure will show rotational symmetry within one full rotation.

7.  6
8.  8
9.  17

Describe the smallest angle of rotational symmetry for each figure.

10.  90°
11.  90°
12.  180°

LESSON **8-11**

Reteach

Symmetry

A figure has **line symmetry** if it can be folded along a line so that the two halves match exactly. The fold line of a figure is called a **line of symmetry**. A figure can have more than one line of symmetry.

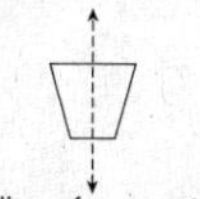

1 line of symmetry

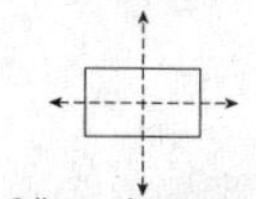

2 lines of symmetry

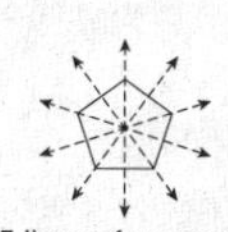

5 lines of symmetry

Trace each figure and cut it out. Fold the figure on each line of symmetry. Tell how many lines of symmetry each figure has.

1. 

1

2.

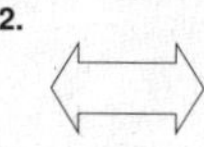

2

3. 

4

A figure has **rotational symmetry** if it can be turned less than a full turn (360°) and look exactly like the original figure.

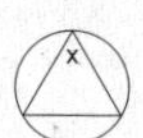

Original figure

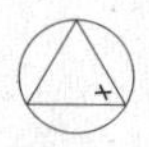

Rotation 1

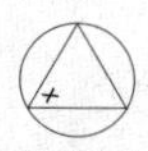

Rotation 2

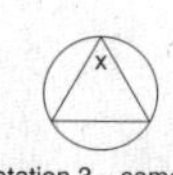

Rotation 3 – same as original figure

Trace each figure. Mark an X at the top of the tracing. Put the tracing on top of the figure and rotate it until it matches. Tell the number of rotations until the X is at the top again.

4. 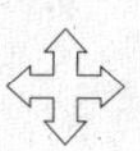

4 rotations

5. 

6 rotations

6. 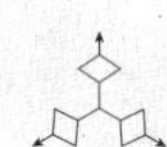

3 rotations

LESSON **8-11**

Challenge

Design Your Own Quilt

This quilt will be a collection of individual squares. In each square, draw a different figure that has the given rotational symmetry.

For example, has $\frac{1}{2}$-turn, or 180°, symmetry.

Answers will vary. Some possible answers given.

$\frac{1}{4}$ turn, or 90°	$\frac{1}{2}$ turn, or 180°	$\frac{1}{4}$ turn, or 90°
$\frac{1}{3}$ turn, or 120°	$\frac{1}{5}$ turn, or 72°	$\frac{1}{3}$ turn, or 120°
$\frac{1}{4}$ turn, or 90°	$\frac{1}{2}$ turn, or 180°	$\frac{1}{4}$ turn, or 90°
$\frac{1}{3}$ turn, or 120°	$\frac{1}{6}$ turn, or 60°	$\frac{1}{3}$ turn, or 120°

LESSON **8-11**

Problem Solving

Symmetry

Write the correct answer.

Use the logo of an Australian TV station for Exercises 1 and 2.

1. How many lines of symmetry do the dots in the Channel 9 logo have?

   4

2. How many times will the dots show rotational symmetry in 1 rotation of 360°? What is the smallest angle of rotational symmetry for the dots?

   4 times; 90°

3. The figure on an emblem will show rotational symmetry 8 times within a single rotation. What is the smallest angle of rotational symmetry for the emblem?

   45°

4. Draw a figure that shows rotational symmetry 5 times within a 360° rotation.

   Possible answer: 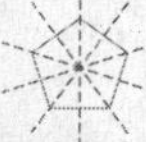

Choose the letter for the best answer.

5. Which is a true statement about the stained glass window?

   A It is asymmetrical.

   (B) It has 1 line of symmetry.

   C It will show rotational symmetry 4 times within a full rotation.

   D It has 2 lines of symmetry.

6. Which is a true statement about the time shown on the clocks?

3:08 4:00 6:13 1:30

1 2 3 4

   F Only clocks 1 and 2 have a line of symmetry.

   G All of the clocks have at least 1 line of symmetry.

   H All of the clocks are asymmetrical.

   (J) Only clocks 1 and 4 have a line of symmetry.

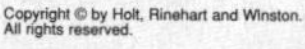

LESSON **8-11**

Reading Strategies

Focus On Example/Non-Example

A figure can rotate around a central point called the **center of rotation.** A figure has **rotational symmetry** if it rotates onto itself before turning one complete turn, or 360°. The center of rotation is a point and is shown with a dot.

If you rotate the letter Z around the central point, it will rotate onto itself halfway around the circle, or at 180°. This figure *has* rotational symmetry

Rotational Symmetry

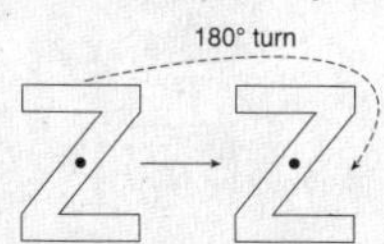

If you rotate the letter E around the central point, it will not rotate onto itself until it has made one full rotation around the circle, at 360°. The figure *does not have* rotational symmetry.

No Rotational Symmetry

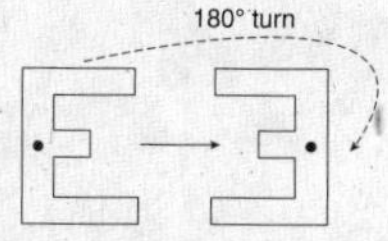

Trace and cut out a copy of each figure. Rotate the cut figure over the same figure on the page. Do these figures have rotational symmetry? Write yes or no.

1. 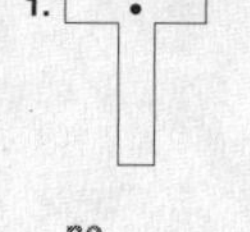

   no

2. 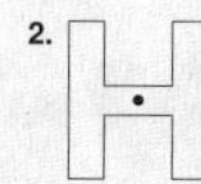

   yes

3.

   no

4. 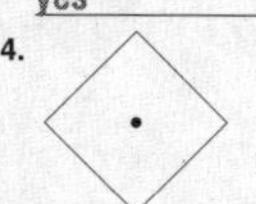

   yes

LESSON 8-11

Puzzles, Twisters & Teasers

This Is Hairy!

Determine the answer for each. Use the letters of your answer to solve the riddle.

1. How many lines of symmetry does this figure have?

(A) 2   **B** 3   **C** 4   **D** None

2. How many lines of symmetry does this figure have?

(O) 1   **A** 2   **E** 3   **H** None

3. How many lines of symmetry does this figure have?

T 2   (L) 6   **M** 16   **R** None

4. How many lines of symmetry does this figure have?

B 2   **S** 3   (T) 4   **H** None

5. How many lines of symmetry does this figure have?

T 2   **S** 3   **T** 4   (P) None

What kind of portable computer does a rabbit like to use?

A	L	O	P	T	O	P
1	3	2	5	4	2	5